Current Topics in Microbiology and Immunology
Volume 345

Current Topics in Microbiology and Immunology

Previously published volumes
Further volumes can be found at springer.com

M. Celeste Simon
Editor

Diverse Effects of Hypoxia on Tumor Progression

 Springer

Editor
M. Celeste Simon, PhD
University of Pennsylvania
School of Medicine
19104-6160 Philadelphia
Pennsylvania
USA
celeste2@mail.med.upenn.edu

ISSN 0070-217X
ISBN: 978-3-642-13328-2   e-ISBN: 978-3-642-13329-9
DOI 10.1007/978-3-642-13329-9
Springer Heidelberg Dordrecht London New York

Library of Congress Control Number: 2010934929

Cover design: WMXDesign GmbH, Heidelberg, Germany

Printed on acid-free paper

Springer is part of Springer Science+Business Media (www.springer.com)

Preface

Solid tumors frequently contain areas of oxygen deprivation (hypoxia) due to rapid cell proliferation and/or vascular insufficiency. The presence of hypoxic domains typically correlates with poor patient prognosis, due to the relative resistance of hypoxic cells to conventional cancer therapies and effects of O_2 availability on disease progression. The response of malignant cells to hypoxia has been the focus of intense research over the last decade. In this issue of *Current Topics in Microbiology and Immunology*, we present chapters describing the impact of hypoxia on components of the tumor microenvironment (such as endothelial cells, inflammatory cells, and tumor-associated fibroblasts), the expression of unique microRNAs, tumor cell differentiation status, and metastasis. Each review chapter describes the state of the field studying these topics and poses important questions for the future. The overall goal is to depict tumor phenotypes and associated molecular pathways to be exploited in the development of novel therapeutics to be used against a broad spectrum of human cancers.

Summer 2010

Celeste Simon

Contents

Contributors

Francesca M. Buffa Molecular Oncology Laboratories, Weatherall Institute of Molecular Medicine, University of Oxford, John Radcliffe Hospital, Oxford OX3 9DS, UK

Peter Carmeliet Vesalius Research Center (VRC), VIB, K.U. Leuven, Campus Gasthuisberg, Herestraat 49, 3000 Leuven, Belgium, peter.carmeliet@vib-kuleuven.be

Amato J. Giaccia Department of Radiation Oncology, Division of Cancer and Radiation Biology, Stanford University School of Medicine, Stanford, CA 94305-5152, USA, giaccia@stanford.edu

Adrian L. Harris Molecular Oncology Laboratories, Weatherall Institute of Molecular Medicine, University of Oxford, John Radcliffe Hospital, Oxford OX3 9DS, UK, aharris.lab@imm.ox.ac.uk

Hongxia Z. Imtiyaz Abramson Family Cancer Research Institute, University of Pennsylvania, 438 BRB II/III, 421 Curie Boulevard, Philadelphia, PA 19104-6160, USA and Howard Hughes Medical Institute, Chevy Chase, MD, USA, hongxiaz@mail.med.upenn.edu

Randall S. Johnson Molecular Biology Section, Division of Biological Sciences, University of California, San Diego, CA 92093, USA, rjohnson@biomail.ucsd.edu

A. Sofie Johnsson Department of Laboratory Medicine, Center for Molecular Pathology, Lund University, University Hospital MAS, Entrance 78, 205 02 Malmo, Sweden

Zhizhong Li Department of Radiation Oncology, Duke University Medical Center, Durham, NC 27710, USA

Robert McCormick Molecular Oncology Laboratories, Weatherall Institute of Molecular Medicine, University of Oxford, John Radcliffe Hospital, Oxford OX3 9DS, UK

Sven Pahlman Department of Laboratory Medicine, Center for Molecular Pathology, Lund University, University Hospital MAS, Entrance 78, 205 02 Malmo, Sweden, sven.pahlman@med.lu.se

Alexander Pietras Department of Laboratory Medicine, Center for Molecular Pathology, Lund University, University Hospital MAS, Entrance 78, 205 02 Malmo, Sweden

Annelies Quaegebeur Vesalius Research Center (VRC), VIB, K.U. Leuven, Campus Gasthuisberg, Herestraat 49, 3000 Leuven, Belgium

Jiannis Ragoussis Molecular Oncology Laboratories, Weatherall Institute of Molecular Medicine, University of Oxford, John Radcliffe Hospital, Oxford OX3 9DS, UK

Jeremy N. Rich Department of Stem Cell Biology and Regenerative Medicine, Lerner Research Institute, Cleveland Clinic, Cleveland, OH 44195, USA, richj@ccf.org

Helene Rundqvist Molecular Biology Section, Division of Biological Sciences, University of California, San Diego, CA 92093, USA

Ernestina Schipani Endocrine Unit, Department of Medicine, MGH-Harvard Medical School, Boston, MA 02114, USA

M. Celeste Simon Abramson Family Cancer Research Institute, University of Pennsylvania, 438 BRB II/III, 421 Curie Boulevard, Philadelphia, PA 19104-6160, USA and Howard Hughes Medical Institute, Chevy Chase, MD, USA and Department of Cell and Developmental Biology, University of Pennsylvania School of Medicine, Philadelphia, PA 19104, USA, celeste2@mail.med.upenn.edu

The HIF-2α-Driven Pseudo-Hypoxic Phenotype in Tumor Aggressiveness, Differentiation, and Vascularization

Alexander Pietras, A. Sofie Johnsson, and Sven Påhlman

Contents

Abstract Cellular adaptation to diminished tissue oxygen tensions, hypoxia, is largely governed by the hypoxia inducible transcription factors, HIF-1 and HIF-2. Tumor hypoxia and high HIF protein levels are frequently associated with aggressive disease. In recent years, high tumor cell levels of HIF-2 and the oxygen sensitive subunit HIF-2α have been associated with unfavorable disease and shown to be highly expressed in tumor stem/initiating cells originating from neuroblastoma and glioma, respectively. In these cells, HIF-2 is active under nonhypoxic conditions as well, creating a pseudo-hypoxic phenotype with clear influence on tumor behavior. Neuroblastoma tumor initiating cells are immature with a neural crest-like phenotype and downregulation of HIF-2α in these cells

A. Pietras, A.S. Johnsson, and S. Påhlman (✉)
Department of Laboratory Medicine, Center for Molecular Pathology, Lund University, University Hospital MAS, Entrance 78, 205 02 Malmö, Sweden
e-mail: sven.pahlman@med.lu.se

M. Celeste Simon (ed.), *Diverse Effects of Hypoxia on Tumor Progression*,
Current Topics in Microbiology and Immunology 345, DOI 10.1007/82_2010_72
© Springer-Verlag Berlin Heidelberg 2010, published online: 2 June 2010

results in neuronal sympathetic differentiation and the cells become phenotypically similar to the bulk of neuroblastoma cells found in clinical specimens. Knockdown of HIF-2α in neuroblastoma and glioma tumor stem/initiating cells leads to reduced levels of VEGF and poorly vascularized, highly necrotic tumors. As high HIF-2α expression further correlates with disseminated disease as demonstrated in neuroblastoma, glioma, and breast carcinoma, we propose that targeting HIF-2α and/or the pseudo-hypoxic phenotype induced by HIF-2 under normoxic conditions has great clinical potential.

1   Introduction

Mammalian cells, including tumor cells, require oxygen for maintenance of an efficient energy supply and lack of oxygen eventually leads to cell death due to impaired energy requiring processes. Cells can withstand fluctuations in oxygen levels by adapting to a decrease in oxygen involving reduction of energy consumption and increase in anaerobic metabolism. During the adaptation process, there is a dramatic shift in the expression of genes regulating a number of cellular functions including glucose transport and metabolism, angiogenesis, and cell survival. Central to this phenotypic shift are the hypoxia inducible factors (HIFs), HIF-1, and HIF-2. These factors are heterodimeric transcription factors composed of a unique alpha subunit and a beta subunit (ARNT/HIF-1β) shared by all three HIFs. Classically, HIFs are regulated by degradation of the alpha subunit at high oxygen levels and by stabilization at hypoxia (reviewed in Kaelin and Ratcliffe 2008). In the dimeric, active state, HIF-1 and HIF-2 bind to hypoxia responsive elements (HREs) located in genes regulated by hypoxia and HIFs. Although HIF-1 and HIF-2 seem to activate hypoxia-responsive genes by similar means (Tian et al. 1997; Wiesener et al. 1998), the HIF-α subunits work in a nonredundant manner and several differences in gene regulation have been proposed, many of which emphasize the predominant role of HIF-1 in regulating the transcriptional response to hypoxia (Iyer et al. 1998; Ryan et al. 1998; Hu et al. 2003; Park et al. 2003; Sowter et al. 2003). As detailed below, HIF-2 is also crucial for the hypoxic response, and as opposed to HIF-1 it is active at prolonged hypoxia as well (Holmquist-Mengelbier et al. 2006). The less studied HIF-3 is present in several splice variants lacking the C-terminal transactivation domain and is thought to negatively regulate HIF-1 and HIF-2 by sequestering the HIF-1 and HIF-2 alpha subunits, thereby blocking their binding to HREs (Makino et al. 2001; Maynard et al. 2007). Although the main mode of HIF activation is via stabilization of the alpha subunits, HIF-2, as well as HIF-1, can be stable and transcriptionally active at physiological or even higher oxygen tensions, as will be the central theme in this review. We propose that this phenomenon, at least regarding HIF-2, is largely linked to its role and regulation during normal development.

The phenotypes obtained by elimination of either *Hif1a* or *Epas1/Hif2a* clearly show that both genes are needed for proper development and that they are nonredundant.

Importantly, and the reason for the focus on HIF-2α in this review, HIF-1α and HIF-2α are differently regulated in tumors such as neuroblastoma, breast cancer, nonsmall cell lung carcinoma (NSCLC), and glioma and seem to have different impact on tumor behavior and patient outcome in these tumors (Holmquist-Mengelbier et al. 2006; Helczynska et al. 2008; Heddleston et al. 2009; Kim et al. 2009b; Li et al. 2009; Noguera et al. 2009).

2 Phenotypic Effects of HIF-2α Elimination

While elimination of *Hif1a* has profound and repeatable effects on embryonal development, the effects of knocking out *Hif2a* have turned out to be much more complex and dependent on the genetic background, as summarized below. Despite displaying incompletely overlapping phenotypes, four different *Hif2a* knockout mice have been instrumental in identifying putative roles for *HIF2A* during normal development (Tian et al. 1998; Peng et al. 2000; Compernolle et al. 2002; Scortegagna et al. 2003). While *Hif1a$^{-/-}$* animals are dead within embryonic day E11 with severe disorganization of vascular networks and gross neural tube defects, the effects of eliminating *Hif2a* – at least during early development – appears less general.

Hif2a expression during development is most abundant in vascular endothelial cells and disrupted vascular development of specific (although distinct) organs has been observed in various *Hif2a$^{-/-}$* mice. Furthermore, whether or not attributable to vascular system defects, some *Hif2a$^{-/-}$* mice have succumbed to embryonic death displaying hemorrhage. In particular, the *Hif2a$^{-/-}$* mice created by Peng et al. showed varying degrees of vascular disorganization despite apparently normal blood vessel formation, suggesting that HIF-2α is required for normal remodeling/maturation postvasculogenesis (2000). *Hif2a$^{-/-}$* mice in other studies, however, appeared normal in vascular development (despite hemorrhage) or displayed only subtle changes during late stages of pulmonary vascularization. In the Compernolle et al. study, *Hif2a* knockout mice died neonatally due to respiratory distress syndrome, apparently caused by impaired fetal lung maturation because of reduced VEGF levels and insufficient surfactant production (2002).

Creating *Hif2a$^{-/-}$* animals by hybrid mating allowed Scortegagna et al. (2003) to study effects of *Hif2a* loss in the postnatal mouse. These mice suffered from biochemical/metabolic abnormalities and multiple-organ pathology specifically in sites of high-energy demand including the heart, liver, testis, and bone marrow, indicating a syndrome related or similar to mitochondrial disease. Overall, adult *Hif2a$^{-/-}$* mice showed greater oxidative stress as well as reduced response to oxidative stress, suggesting an important role for HIF-2α in ROS homeostasis. The fact that *Hif2a* itself is regulated by ROS may indicate a role as a primary sensor of oxidative stress. In support, ROS accumulation and improved response to radiation therapy by *HIF2A* inhibition was recently described in human tumor cells

(Bertout et al. 2009). These findings may implicate a role for *HIF2A* in radiation and chemotherapy resistance in tumor and possibly normal stem cells.

In addition, $Hif2a^{-/-}$ mice display defects in hematopoietic development due to greatly reduced EPO levels in the kidney (Scortegagna et al. 2003, 2005; Rankin et al. 2007). Administration of exogenous EPO reverts this phenotype as well as some of the other defects associated with *Hif2a* elimination (Scortegagna et al. 2005). Further supporting a role for *HIF2A* in EPO production, a gain-of-function mutation in the *HIF2A* gene has been found associated with familial erythrocytosis (Percy et al. 2008a, b).

3 HIF-2 During Normal Sympathetic Nervous System (SNS) Development

In 1998, Steven McKnight and colleagues showed that *Hif2a* expression was transient but prominent in developing sympathetic ganglia and paraganglia (Organ of Zuckerkandl) (Tian et al. 1998). The latter organ is the main site of catecholamine synthesis during development and is thus tyrosine hydroxylase (TH) positive; it was later confirmed that HIF-2α is also expressed in developing human fetal SNS (Nilsson et al. 2005) (Fig. 1). Supporting a direct role for *Hif2a* in catecholamine production, the *Hif2a* deficient 129/SvJ mice contained substantially reduced levels of catecholamines, displayed bradycardia, and died at mid-gestation at a developmental stage corresponding to when *Hif2a* levels in the Organ of

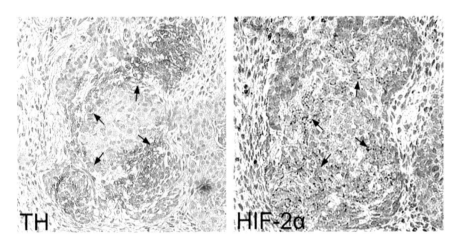

Fig. 1 *HIF-2α positive human paraganglia at fetal week 8.5.* Paraganglia stained for tyrosine hydroxylase (TH) and HIF-2α in nonconsecutive but adjacent sections (ethical approval LU 389-98, Lund University, Sweden). *Arrows* indicate distinct nests of immature paraganglia cells positive for both TH and HIF-2α. These structures have been further characterized in (Hoehner et al. 1996)

Zuckerkandl were the highest in heterozygous animals (Tian et al. 1998). Strikingly, the mid-gestational death was rescued by feeding the mothers DOPS, a substance that can directly convert into norepinephrine. Although a sympathetic phenotype was less pronounced in other $Hif2a^{-/-}$ mice – particularly as a cause of death – altered catecholamine content or DOPS-mediated rescue was recorded at least to some degree in all other $Hif2a$ knockout animals (Peng et al. 2000; Compernolle et al. 2002; Scortegagna et al. 2003). These findings are consistent with the reported role of $Hif2a$ in activating transcription of the DDC and DBH enzymes and thereby regulating catecholamine synthesis in fetal rat sympathoadrenal progenitor cells regardless of oxygen tension (Brown et al. 2009). In further support of a role for $Hif2a$ in sympathetic development, mice lacking the HIF prolyl hydroxylase PHD3 displayed an increased number (but reduced functionality) of sympathetic cells in the adrenal medulla, carotid body, and the superior cervical ganglia due to reduced apoptosis (Bishop et al. 2008). A reasonable assumption based on the role of PHD3 in targeting HIFs for degradation is that HIF protein levels in general would be higher in $PHD3^{-/-}$ animals, and *in vitro* studies have suggested that PHD3 is more important in regulation of HIF-2α than HIF-1α (Appelhoff et al. 2004; Henze et al. 2009). Indeed, the sympathetic phenotype of $Phd3^{-/-}$ mice was intriguingly reverted by crossing animals with heterozygous $Hif2a^{+/-}$ (but not $Hif1a^{+/-}$) mice, again indicating that proper control of $Hif2a$ expression is crucial for normal SNS development (Bishop et al. 2008). This notion is perhaps embodied by the link between high HIF-2α expression and immature, aggressive phenotypes of the SNS malignancy neuroblastoma, as discussed below.

4 Hypoxia in Solid Tumors and Relation to Tumor Aggressiveness

Direct measurements of oxygen tension in solid tumors and adjacent nonmalignant tissue reveal that tumors, generally, are less well oxygenated and that large parts of solid tumors are hypoxic (Höckel and Vaupel 2001). Although these hypoxic areas are often necrotic, the general histological pattern is such that tumor cells survive low oxygen tensions and can thus adapt to hypoxic conditions. *In vitro* studies support this conclusion as tumor cells established as cell lines can survive for several days at as low concentrations as 0.1% oxygen. Another interesting aspect of tumor hypoxia is the well-documented association between oxygen shortage and tumor aggressiveness (reviewed in Bertout et al. 2008). The mechanistic background is probably very complex, but involves cytotoxic resistance, insensitivity to radiation, decreased DNA repair capacity, increased vascularization, and increased metastatic potential (reviewed in Semenza 2003; Erler et al. 2006; Löfstedt et al. 2007) and as will be discussed in more detail below, dedifferentiation or loss of a differentiated tumor phenotype. As adaptation to hypoxia in tumor cells is largely mediated via stabilization and activation of HIF-1α and HIF-2α, high levels of HIF proteins have also been associated with disseminated disease and poor overall survival. In tumor cell lines – with few exceptions – both HIF-1α and HIF-2α are expressed and we recently postulated that HIF-1α is involved in adaptation to acute and HIF-2α to prolonged hypoxia (Holmquist-Mengelbier et al. 2006; Helczynska et al. 2008). For historical reasons, HIF-1α is the isoform that has been most extensively studied in clinical tumor materials and frequently been correlated to aggressive tumor disease, but in recent years high tumor levels of HIF-2α rather than HIF-1α have been shown to associate with negative overall survival and metastatic disease. In breast carcinoma for instance, earlier published data link HIF-1α, while later reports link HIF-2α to unfavorable disease (Schindl et al. 2002; Bos et al. 2003; Gruber et al. 2004; Dales et al. 2005; Generali et al. 2006; Giatromanolaki et al. 2006; Kronblad et al. 2006; Helczynska et al. 2008). Whether these contradicting observations reflect real differences in the tumor material analyzed or can be attributed to methodological shortcomings is presently unknown. Nevertheless, data from tumors of different derivations in which HIF-2 appears to be important for clinical behavior are exemplified in the next paragraph.

5 Differential Tumor HIF Expression in Relation to Patient Outcome

5.1 Neuroblastoma

Neuroblastoma is a childhood tumor that arises in precursor cells or immature neuroblasts of the SNS, which is derived from the neural crest. There is strong positive correlation between tumor aggressiveness (clinical stage and overall outcome) and immature phenotype (Fredlund et al. 2008). Although several cytogenetic aberrations linked to poor neuroblastoma prognosis have been identified, amplification of MYCN is the only aberration at the gene level that strongly associates with advanced disease, which in turn associates with activated MYC signaling and an immature phenotype (Fredlund et al. 2008). The fully disseminated disease, Stage 4, is highly aggressive and the overall survival of children with this disease stage is less than 40% (Matthay et al. 1999).

   As stated above, HIF-2α is expressed during discrete periods of murine and human SNS development and the fact that neuroblastoma is an SNS-derived tumor appears to be important when the role(s) of HIFs in neuroblastoma is discussed. Both HIF-1α and HIF-2α proteins are expressed and become stabilized at hypoxia in neuroblastoma cell lines (Jögi et al. 2002). There is however a distinct difference in stabilization kinetics suggesting that HIF-1 is responsible for the acute, and HIF-2 for the prolonged response to hypoxia (Holmquist-Mengelbier et al. 2006). HIF-2α is also less sensitive than HIF-1α to oxygen-dependent degradation, and accumulates already at near-physiological oxygen tensions. Immunohistochemical analysis of HIF expression in neuroblastoma specimens reveals that both HIF-1α and HIF-2α proteins, as expected, can be detected in tumor cell layers adjacent to necrotic areas. While HIF-1α is mainly restricted to perinecrotic zones, HIF-2α protein is also expressed at other locations, most notably in cells adjacent to blood vessels. Presence of tumor cells staining intensely for HIF-2α, more so than high number of HIF-2α$^+$ cells correlates positively to distant metastasis and negative overall survival (Holmquist-Mengelbier et al. 2006; Noguera et al. 2009). In contrast, HIF-1α protein expression did not correlate to aggressive disease or negative outcome (Noguera et al. 2009). As will be discussed below, we postulated that a fraction of the cells staining intensely for HIF-2α are the neuroblastoma tumor initiating or stem cells, which could explain why the presence of such cells so strongly associates with unfavorable disease (Pietras et al. 2008).

5.2 Breast Carcinoma

HIF-1α protein is not expressed in normal breast tissue or ductal hyperplastic lesions, but is detected in ductal carcinoma in situ (DCIS) and invasive breast cancers (Bos et al. 2001; Helczynska et al. 2003) In cell lines, both HIFs are expressed and hypoxia-regulated. Similar to the situation in neuroblastoma, HIF-1α is acutely and transiently upregulated, whereas HIF-2α protein is still present after prolonged hypoxia and appears to mediate a sustained hypoxic response, including expression of VEGF (Helczynska et al. 2008). Expression in tumors and association of HIFs to breast cancer aggressiveness appear to be a complex issue. Early studies on HIF-1α protein expression in various subgroups of breast cancers link high levels of the protein to poor outcome, although several of these reports contradict each other. In more recent studies, the overall relationship between HIF-1α protein and breast cancer specific death is meager and HIF-1α associates positively rather than negatively to favorable disease (Tan et al. 2007; Helczynska et al. 2008). However, there are early reports correlating high HIF-1α protein expression with shorter overall and disease-free survival time in patients with lymph node-positive breast cancer, whereas this association was not significant in lymph-node negative patients (Schindl et al. 2002; Kronblad et al. 2006). In contrast to these findings, two reports show association between high HIF-1α protein expression and poor outcome in node-negative but not in node-positive subgroups of patients (Bos et al. 2003; Generali et al. 2006). In addition, significant associations between HIF-1α protein expression and outcome without subgroup divisions (Dales et al. 2005) and unfavorable outcome in node-positive tumors, although restricted to T1/T2 tumors (Gruber et al. 2004) have been published. There are several putative explanations as to why these reports differ in predicting outcome and range from small or poorly defined clinical material to technical explanations. Our own experience is that commercial HIF antibodies vary in quality, also at the batch level, implicating that immunohistochemical stainings have to be interpreted with some caution. In summary, we conclude from published data that the prognostic impact of HIF-1α protein expression in breast cancer is at best restricted to subgroups of patients, which in such cases need to be verified in large prospective studies. Most studies, however, have common findings in that multi- and univariate analyses fail to reveal HIF-1α protein level as an independent prognostic factor.

HIF-2α and its association with the outcome in breast cancer patients has been far less studied, but published immunohistochemical data suggest that HIF-2α correlates to high metastatic potential and is an independent prognostic factor associated with breast cancer specific death (Helczynska et al. 2008). In two cohorts of breast cancer patients, both HIF-1α and HIF-2α correlated to increased VEGF expression, but only high HIF-2α protein exhibited significant correlation to reduced recurrence-free and breast cancer-specific survival, and was an independent prognostic factor. Importantly, high HIF-2α protein expression correlated to the presence of distal metastasis but to no other clinical feature analyzed (Helczynska et al. 2008). In another report, HIF-2α protein was analyzed in a small subset of infiltrating ductal breast carcinomas, which showed a significant relationship between high HIF-2α protein expression and increased vascular density as well as secondary deposits to multiple axillary lymph nodes. Multivariate analysis revealed HIF-2α as an independent factor relating to extensive nodal metastasis (Giatromanolaki et al. 2006).

5.3 Renal Cell Carcinoma

During normal kidney development, HIF-1α is expressed in most cell types whereas HIF-2α is mainly found in renal interstitial fibroblast-like cells and endothelial cells. In the fully developed normal kidney, HIF-1α expression is maintained, while HIF-2α expression disappears. The role of HIF-signaling during development is largely unclear, but the cell type- and stage-specific expression distribution of HIF-α subunits correlates with the expression of critical angiogenic factors such as VEGF and endoglin (Freeburg and Abrahamson 2003; Bernhardt et al. 2006). Conditional knockouts in renal proximal tubule cells of either HIF-1α or HIF-β alone do not generate an abnormal phenotype whereas conditional knockout of pVHL results in HIF-dependent development of tubular and glomerular cysts (Rankin et al. 2006).

Clear cell renal cell carcinoma (CCRCC) is characterized by extensive neovascularization. This is generally explained by impaired HIF-α subunit degradation due to mutation or hypermethylation of the *VHL* gene, found in approximately 60–70% of all CCRCCs (Gnarra et al. 1994; Herman et al. 1994). At normoxia, pVHL constitutes the recognition subunit of a larger E3 ubiquitin ligase complex that targets the HIF-α subunits for proteasomal degradation (Kaelin 2002). Thus, in CCRCCs where pVHL function has been lost the HIF-α subunits are constitutively expressed and a pseudohypoxic phenotype, including increased vascularization, is present. Intriguingly, there seems to be a bias towards HIF-2α expression as compared to HIF-1α expression in these *VHL*-deficient carcinoma cells (Maxwell et al. 1999; Krieg et al. 2000).

The abundance of *VHL*-deficient RCC cell lines expressing HIF-2α but not HIF-1α (Maxwell et al. 1999) is also interesting as this contrasts with normal renal epithelial cells, where HIF-2α expression is absent during ischemia (Rosenberger et al. 2003). Furthermore, the HIF signaling pathways are activated early in the development of neoplastic lesions in VHL disease, with the HIF-1α isoform being expressed even in earliest foci while the HIF-2α protein is detected first in more advanced lesions (Mandriota et al. 2002). In pVHL-defective CCRCC, HIF-1 positively regulates BNIP3, an autophagy marker, but has no profound effect on cyclin D1, TGF-α, and VEGF expression, whereas HIF-2 negatively regulates BNIP3 but promotes cyclin D1, TGF-α, and VEGF expression (Raval et al. 2005). Thus, these differences in regulation of autophagy vs. cell growth and angiogenesis might be understood in the light of HIF-2α being expressed mainly during late CCRCC progression and in more advanced lesions. siRNA-mediated knockdown of HIF-2α represses tumor growth in pVHL-deficient CCRCC (Kondo et al. 2003; Zimmer et al. 2004), and overexpression of HIF-2α in the *VHL* wild type 786-O cells resulted in enhanced tumor formation (Raval et al. 2005). In contrast, overexpression of HIF-1α in 786-O cells diminished tumor xenograft growth (Raval et al. 2005). Finally, and in agreement with the HIF-1α overexpression xenograft data, HIF-1α has been reported in a clinical RCC material to be an independent prognostic factor predicting favorable outcome (Lidgren et al. 2005).

5.4 Nonsmall Cell Lung Carcinoma

HIF protein expression is virtually absent in normal lung tissue at normoxia, whereas both isoforms are accumulated during hypoxic conditions (Giatromanolaki et al. 2001). In the corresponding normal lung tissue examined from lung cancer patients, bronchial and alveolar epithelium adjacent to the tumor site show weak to intense cytoplasmic staining of the HIF proteins, whereas all other lung tissue components are negative for HIF expression (Giatromanolaki et al. 2001).

Intratumoral hypoxia in lung cancers correlates with decreased overall survival (Swinson et al. 2003; Le et al. 2006). Both HIF-1α and HIF-2α are frequently expressed in NSCLC, also during early progression of disease, but whereas HIF-2α causes, or is a surrogate marker for poor clinical prognosis (Giatromanolaki et al. 2001), the role of HIF-1α in predicting outcome is debated. Some reports demonstrate that HIF-1α expression has no impact on patient overall survival (Giatromanolaki et al. 2001; Kim et al. 2005), but potentially contradicting data exist (Volm and Koomagi 2000; Yohena et al. 2009).

In lung adenocarcinomas, mutations in *KRAS* are common and the presence of *KRAS* mutations predicts poor outcome (Huncharek et al. 1999). Mice conditionally expressing a nondegradable HIF-2α and mutated *Kras* ($Kras^{G12D}$) in the lungs display severed tumor burden and decreased survival, compared to mice expressing $Kras^{G12D}$ only, suggesting that HIF-2α play a pivotal role in lung cancer pathogenesis (Kim et al. 2009b). In agreement with a role for HIF-2α in NSCLC tumorigenesis, in a clinical material HIF-2α was an independent prognostic marker with high protein expression correlating to poor outcome (Giatromanolaki et al. 2001).

5.5 Glioblastoma

Glioblastoma multiforme (GBM) is characterized by a rich vasculature network (Hossman and Bloink 1981; Blasberg et al. 1983; Groothuis et al. 1983) and intratumoral necrosis (Raza et al. 2002). Both HIF-1α and HIF-2α proteins are expressed in human glioblastomas (Jensen 2006; Li et al. 2009) with HIF-1α expression being mostly concentrated in areas of necrosis and at the tumor margin (Zagzag et al. 2000). Studies on a small set of brain tumors have suggested that HIF-1α protein correlates positively to brain tumor grade and vascularity (Zagzag et al. 2000).

Based on published data, the role of HIF-2α in glioblastoma formation and aggressiveness is not fully clear, as HIF-2α has been attributed a tumor-suppressor role (Acker et al. 2005), as well as a marker for poor prognosis (Li et al. 2009). Overexpression of HIF-2α protein in rat glioblastomas suppressed tumor growth despite overall enhanced vascularization. This was in part explained by increased tumor cell apoptosis, and knockdown of HIF-2α in hypoxic human glioblastoma cells reduced the apoptotic rate of these cells (Acker et al. 2005). Recent work on GBM has focused highly on the small fraction of tumor cells with stem cell characteristics that are thought to initiate and maintain tumor growth (Hemmati et al. 2003; Singh et al. 2003; Galli et al. 2004; Singh et al. 2004). Several markers identifying a glioma stem cell population have been proposed, including CD133, Nestin (Singh et al. 2003), and A2B5 (Ogden et al. 2008). HIF-2α was recently shown to be expressed at high levels in CD133[+] glioma stem cells grown in vitro (McCord et al. 2009), and to co-localize with stem cell markers in tumor specimens (Li et al. 2009), suggesting that HIF-2α is an independent marker for glioma stem cells. Interestingly, HIF-2α is specifically expressed in brain tumor stem cells but not in neural progenitor cells, in contrast to HIF-1α, which is expressed in both cell types. As in neuroblastoma, a proportion of the HIF-2α positive cells are located adjacent to blood vessels in the tumor specimens, indicating that HIF-2α is expressed by a small but significant number of tumor cells, also in nonhypoxic regions (Pietras et al. 2008, 2009; Li et al. 2009). Finally, analyzing gliomas at the mRNA level, HIF2A, but not HIF1A expression, correlates with poor patient survival (Li et al. 2009).

6 Hypoxia and Tumor Cell Differentiation

As mentioned above, hypoxia has profound effects on cellular phenotypes. One aspect of adaptation to hypoxia, which is of particular importance in tumor cells, is the effect on the tumor cell differentiation status and newly discovered links between HIF-2α expression and tumor initiating/stem cells. Initially described in cultured neuroblastoma and breast cancer cells and in breast tumor specimens, hypoxia can push tumor cells towards an immature stem cell-like phenotype (Jögi et al. 2002; Helczynska et al. 2003). The phenomenon has also been observed in glioma (Heddleston et al. 2009) recently suggesting that the dedifferentiating effect of hypoxia could be general and not restricted to specific tumor forms. These observations have potential direct clinical impact, since at least in neuroblastoma and breast carcinoma, immature stages of differentiation correlate to aggressive tumor behavior and unfavorable outcome. Thus, we have proposed that the hypoxia-induced immature stem cell features work in concert with other hypoxia-driven changes in establishing an aggressive tumor phenotype (Jögi et al. 2002; Helczynska et al. 2003; Axelson et al. 2005).

7 HIF-2α and Tumor Initiating/Stem Cells

HIF-2α is expressed during discrete periods of murine SNS development as determined by in situ hybridization (Tian et al. 1998) and the expression is both strong and selective as most other tissues either lack or only show week HIF-2α expression (Jögi et al. 2002), suggesting that HIF-2α in the developing SNS is regulated at the transcriptional level. By immunohistochemistry we could further demonstrate HIF-2α protein in human SNS paraganglia at fetal week 8.5 (Nilsson et al. 2005), which developmentally corresponds to mouse embryonal day E16, a time point when mouse SNS paraganglia express HIF-2α as determined by in situ hybridizations (Tian et al. 1998; Jögi et al. 2002). Using the same immunohistochemical protocol that detects HIF-2α in developing human paraganglia, staining of human neuroblastoma specimens highlights small subsets of cells intensely expressing HIF-2α protein, and the presence of such cells strongly correlates to disseminated disease (high clinical stage) and tumor related death (Holmquist-Mengelbier et al. 2006). Further immunohistochemical characterization of these cells reveals that they are frequently perivascularly located, lack the expression of SNS markers like TH and NSE found in the bulk of neuroblastoma tumor cells, but express neural crest and early SNS progenitor markers such as NOTCH-1, HES-1, Vimentin, and HAND2 (Pietras et al. 2008). Histologically, these cells were classified as tumor cells, although ambiguous cases exist. However, in most cases it could be excluded that the HIF-2α$^+$ cells were tumor-associated macrophages, reported to express HIF-2α, and contributed to adverse outcome when present in breast cancer specimens (Leek et al. 2002). To verify that the HIF-2α$^+$ cells indeed were tumor cells proper and not stromal cells, MYCN amplification was demonstrated by in situ FISH in perivascularly located, strongly HIF-2α immunofluorescing cells in tumors harboring an amplified MYCN gene. We hypothesized that these immature stem cell-like HIF-2α$^+$ cells could be neuroblastoma stem or tumor initiating cells (TICs) (Pietras et al. 2008).

Recently, David Kaplan's laboratory isolated neuroblastoma cells from patient bone marrows and showed that these cells grow and form neurospheres in neural stem cell promoting medium (Hansford et al. 2007). These cells are highly tumorigenic in an orthotopic xenograft mouse model (Hansford et al. 2007) and are by this functional definition TICs. The neuroblastoma TICs virtually lack expression of SNS markers but express neural crest markers including *NOTCH1*, *HES1*, *ID2*, and *VIM* (Pietras et al. 2009). As the TICs also have high levels of HIF-2α at normoxic conditions, they strongly share phenotypic characteristics with the earlier identified HIF-2α^+, SNS marker$^-$, and neural crest marker$^+$ cells in neuroblastoma specimens (Pietras et al. 2008). The relation between isolated immature neuroblastoma bone marrow TICs and the phenotypically similar cells in neuroblastoma specimens, and that between immature stem cell-like cells in tumor specimens and the bulk of neuroblastoma cells expressing SNS markers have not been established. However, down-regulation of HIF-2α in the cultured neuroblastoma TICs by an sh*HIF2A* approach releases the tumor cells from a differentiation block resulting in expression of the early SNS markers *ASCL1/HASH1*, *ISL1*, and *SCG10*. When removed from the stem cell-promoting medium and grown *in vivo* as subcutaneous tumors, the shHIF-2α-transduced TICs develop into a more mature neuroblastoma phenotype with expression of classical SNS markers such as tyrosine hydroxylase and chromogranin A, thus acquiring a phenotype similar to that of the bulk cells of clinical neuroblastomas (Pietras et al. 2009). We conclude that HIF-2α keeps the neuroblastoma TICs in a stem cell-like state and that these cells have properties in keeping with what could be expected of a neuroblastoma stem cell. Our current view of the relation between neuroblastoma TICs, circulating neuroblastoma cells, tumor bulk, and HIF protein expression is summarized in Fig. 2. The phenotypic similarities between bone marrow-derived TICs and the HIF-2α^+ tumor cells located adjacent to blood vessels in neuroblastoma specimens suggest that these cells are related and we postulate that circulating neuroblastoma cells are the connecting link as has been demonstrated in melanoma, breast, and colon tumor model systems by Massague and co-workers (Kim et al. 2009a). In tumors, we further postulate that the HIF-2α^+ neuroblastoma stem cells will by unknown mechanisms spontaneously differentiate bulk cells expressing SNS markers such as *CHGA*, *TH*, and *GAP-43*. The bulk cells have reduced normoxic VEGF expression and thus reduced angiogenic capacity due to lowered HIF-2α protein levels and when such cells experience hypoxia, they dedifferentiate and acquire stem cell neural crest-like features (Jögi et al. 2002).

As briefly touched upon above, there is also experimental support from other cell systems for the role of HIF-2α during early development and maintenance of a (tumor) stem cell phenotype (reviewed in Keith and Simon 2007). In embryoid bodies, overexpression of HIF-2α results in maintained pluripotency and potentiation of tumorigenic growth (Covello et al. 2006). These effects were a result of direct transcriptional activation of the POU transcription factor *OCT4* by HIF-2α, as silencing of *OCT4* in HIF-2α knock-in cells reverted the stem cell phenotype and reduced tumor growth (Covello et al. 2006). In glioma, cell populations enriched for tumor stem cell properties have high HIF-2α protein levels and as in

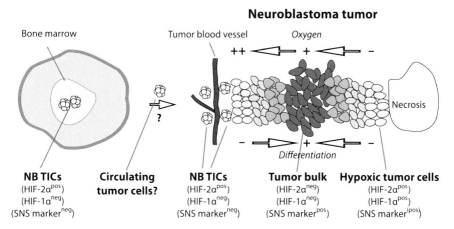

Fig. 2 *A putative interplay between neuroblastoma (NB) tumor-initiating cells (TICs), tumor bulk, HIF-2α, sympathetic nervous system (SNS) differentiation, and oxygen status.* We postulate that bone marrow-derived neuroblastoma TICs communicate with primary neuroblastomas as circulating tumor cells. In neuroblastoma tumors, TICs will by unknown mechanisms, spontaneously lose their HIF-2α protein expression, differentiate, and acquire expression of SNS markers. In hypoxic regions of neuroblastomas, tumor cells lose their differentiated phenotype and become stem cell-like (Jögi et al. 2002). In this model, HIF-1α protein expression is strictly linked to a hypoxic cellular environment

neuroblastoma, HIF-2α has been suggested to be a marker of glioma stem cells. In addition, downregulation of HIF-2α in such cells results in decreased tumor initiating capacity and glial differentiation (Li et al. 2009; Heddleston et al. 2009). In line with these findings are the observations that HIF-2α protein expression correlates to breast cancer specific death and distant metastasis – the latter process most likely dependent on the presence of cells with tumor stem cell properties. Based on the published observations that HIF-2α is intimately and functionally linked to immature neural tumor stem cell phenotypes and appears to counteract early steps in SNS and glial differentiation, we hypothesize that HIF-2α might be a general marker of tumor stem cells with a dedifferentiating function similar to that in glioma and neuroblastoma stem/TICs.

8 HIFs and Vascularization

HIFs were implicated early in the tumor angiogenic process when it became clear that hypoxia promotes VEGF expression (Shweiki et al. 1992; Forsythe et al. 1996). With neuroblastoma as a model system we showed that there is a temporal shift in the usage of the HIFs during hypoxia-driven VEGF expression; whereas the VEGF expression is HIF-1 dependent at an acute phase, the expression during prolonged hypoxia is primarily HIF-2 driven (Holmquist-Mengelbier et al. 2006). In a follow-up study using large clinical neuroblastoma tissue microarray material immunohisto-chemically stained for HIF-1α, HIF-2α, VEGF, and blood vessel endothelial cells (CD31), tumor cells staining intensely for HIFs correlated to VEGF positivity, while the HIF-1α and HIF-2α staining did not fully correlate to each other (Noguera et al. 2009). Furthermore, HIF-1α and VEGF, but not HIF-2α, showed negative correlation to the number of blood vessels, in agreement with the observation that strongly HIF-2α positive tumor cells express VEGF and frequently locate adjacent to blood vessels. As opposed to HIF-2α, presence of HIF-1α positive cells did not correlate to tumor aggressiveness or disseminated disease (Noguera et al. 2009). We conclude from *in vitro* and *in vivo* neuroblastoma data that both HIF-1α and HIF-2α become stabilized at hypoxia and that the HIF-2α protein level can be positively regulated by additional, not fully explored mechanisms.

The HIF-2α$^+$ neuroblastoma and glioma stem/TICs strongly express VEGF (Holmquist-Mengelbier et al. 2006; Bao et al. 2006; Calabrese et al. 2007; Pietras et al. 2008, 2009; Li et al. 2009) and the perivascular location of phenotypically similar cells in tumor specimens might implicate that tumor stem cells actively take part in the process of tumor vascularization. The hypothesis is supported by the observation that knockdown of HIF-2α in neuroblastoma and glioma stem/TICs results in poorly vascularized tumors. Moreover, vascularization is affected in mice with eliminated *Hif2a* (Peng et al. 2000; Rankin et al. 2008) and overexpression of *Hif2a* in embryoid bodies results in early and extensive formation of a vascular network (Covello et al. 2006). The underlying molecular mechanisms behind the perivascular localization of neural tumor stem cells are presently not understood and are topics for future investigations.

9 HIF-2α and the Pseudo-Hypoxic Phenotype: Targets for Tumor Treatment

One striking feature of both neuroblastoma and glioma stem/TICs is their pseudo-hypoxic phenotype due to high expression of active HIF-2α at physiological oxygen tensions (Holmquist-Mengelbier et al. 2006; Pietras et al. 2008, 2009; Li et al. 2009). As a consequence, several genes considered to be hypoxia-regulated are highly expressed at nonhypoxic conditions in these HIF-2α$^+$ tumor cells, thus creating a pseudo-hypoxic phenotype, presumably similar to that in a majority of CCRCCs. Whether the pseudo-hypoxic phenotype is drastically different from a corresponding *bona fide* hypoxic phenotype is not known, but as discussed above, VEGF expression and presumably vascularization would be one important deter-minant of the high HIF-2α expression at nonhypoxic conditions. This conclusion, in addition to data indicating that HIF-2 appears to play a pivotal role in aggressive and disseminated growth of neuroblastoma, glioma, breast carcinoma, nonsmall cell lung carcinoma, and CCRCC, strongly suggests HIF-2α and the pseudo-hypoxic phenotype are attractive therapeutic targets in at least these tumor types.

By pushing HIF-2α⁺ tumor stem cells toward a more mature tumor bulk-like phenotype by interfering with the expression or activity of HIF-2, the tumor stem cell pool as well as the bulk of tumor cells, targeted by existing, efficient treatment protocols, would be reduced in numbers. The demonstration that both neuroblastoma and glioma stem/initiating cells differentiate *in vivo* when HIF-2α is knocked down (Pietras et al. 2009; Heddleston et al. 2009; Li et al. 2009) can be seen as proof of principle. Diminished HIF-2α activity also leads to reduced VEGF expression in these two tumor stem cell models, and as would be anticipated, neuroblastoma and glioma tumors with reduced HIF-2α expression are highly necrotic (Pietras et al. 2009; Li et al. 2009; Heddleston et al. 2009) suggesting that targeting HIF-2α will result in both an antiangiogenic effect and a reduction of the tumor stem cell pool. As high HIF-2α protein levels associate with disseminated disease, targeting of HIF-2α or the HIF-2α-driven pseudo-hypoxic phenotype, might also affect tumor spread and transition into advanced clinical stages.

Acknowledgments This work was supported by the Swedish Cancer Society, the Children's Cancer Foundation of Sweden, the Swedish Research Council, the SSF Strategic Center for Translational Cancer Research – CREATE Health, Gunnar Nilsson's Cancer Foundation and the research funds of Malmö University Hospital.

References

Acker T, Diez-Juan A, Aragones J, Tjwa M, Brusselmans K, Moons L, Fukumura D, Moreno-Murciano MP, Herbert JM, Burger A, Riedel J, Elvert G, Flamme I, Maxwell PH, Collen D, Dewerchin M, Jain RK, Plate KH, Carmeliet P (2005) Genetic evidence for a tumor suppressor role of HIF-2alpha. Cancer Cell 8:131–141

Appelhoff RJ, Tian YM, Raval RR, Turley H, Harris AL, Pugh CW, Ratcliffe PJ, Gleadle JM (2004) Differential function of the prolyl hydroxylases PHD1, PHD2, and PHD3 in the regulation of hypoxia-inducible factor. J Biol Chem 279:38458–38465

Axelson H, Fredlund E, Ovenberger M, Landberg G, Påhlman S (2005) Hypoxia-induced dedifferentiation of tumor cells–a mechanism behind heterogeneity and aggressiveness of solid tumors. Semin Cell Dev Biol 16:554–563

Bao S, Wu Q, Sathornsumetee S, Hao Y, Li Z, Hjelmeland AB, Shi Q, McLendon RE, Bigner DD, Rich JN (2006) Stem cell-like glioma cells promote tumor angiogenesis through vascular endothelial growth factor. Cancer Res 66:7843–7848

Bernhardt WM, Schmitt R, Rosenberger C, Munchenhagen PM, Grone HJ, Frei U, Warnecke C, Bachmann S, Wiesener MS, Willam C, Eckardt KU (2006) Expression of hypoxia-inducible transcription factors in developing human and rat kidneys. Kidney Int 69:114–122

Bertout JA, Patel SA, Simon MC (2008) The impact of O2 availability on human cancer. Nat Rev Cancer 8:967–975

Bertout JA, Majmundar AJ, Gordan JD, Lam JC, Ditsworth D, Keith B, Brown EJ, Nathanson KL, Simon MC (2009) HIF2alpha inhibition promotes p53 pathway activity, tumor cell death, and radiation responses. Proc Natl Acad Sci USA 106:14391–14396

Bishop T, Gallagher D, Pascual A, Lygate CA, de Bono JP, Nicholls LG, Ortega-Saenz P, Oster H, Wijeyekoon B, Sutherland AI, Grosfeld A, Aragones J, Schneider M, van Geyte K, Teixeira D, Diez-Juan A, Lopez-Barneo J, Channon KM, Maxwell PH, Pugh CW, Davies AM, Carmeliet P, Ratcliffe PJ (2008) Abnormal sympathoadrenal development and systemic hypotension in PHD3-/- mice. Mol Cell Biol 28:3386–3400

Blasberg RG, Kobayashi T, Horowitz M, Rice JM, Groothuis D, Molnar P, Fenstermacher JD
(1983) Regional blood flow in ethylnitrosourea-induced brain tumors. Ann Neurol 14:189–201

Bos R, Zhong H, Hanrahan CF, Mommers EC, Semenza GL, Pinedo HM, Abeloff MD, Simons JW,
van Diest PJ, van der Wall E (2001) Levels of hypoxia-inducible factor-1 alpha during breast
carcinogenesis. J Natl Cancer Inst 93:309–314

Bos R, van der Groep P, Greijer AE, Shvarts A, Meijer S, Pinedo HM, Semenza GL, van Diest PJ,
van der Wall E (2003) Levels of hypoxia-inducible factor-1alpha independently predict
prognosis in patients with lymph node negative breast carcinoma. Cancer 97:1573–1581

Brown ST, Kelly KF, Daniel JM, Nurse CA (2009) Hypoxia inducible factor (HIF)-2 alpha is
required for the development of the catecholaminergic phenotype of sympathoadrenal cells.
J Neurochem 110:622–630

Calabrese C, Poppleton H, Kocak M, Hogg TL, Fuller C, Hamner B, Oh EY, Gaber MW,
Finklestein D, Allen M, Frank A, Bayazitov IT, Zakharenko SS, Gajjar A, Davidoff A,
Gilbertson RJ (2007) A perivascular niche for brain tumor stem cells. Cancer Cell 11:69–82

Compernolle V, Brusselmans K, Acker T, Hoet P, Tjwa M, Beck H, Plaisance S, Dor Y, Keshet E,
Lupu F, Nemery B, Dewerchin M, Van Veldhoven P, Plate K, Moons L, Collen D, Carmeliet P
(2002) Loss of HIF-2alpha and inhibition of VEGF impair fetal lung maturation, whereas treatment with VEGF prevents fatal respiratory distress in premature mice. Nat Med
8:702–710

Covello KL, Kehler J, Yu H, Gordan JD, Arsham AM, Hu CJ, Labosky PA, Simon MC, Keith B
(2006) HIF-2alpha regulates Oct-4: effects of hypoxia on stem cell function, embryonic
development, and tumor growth. Genes Dev 20:557–570

Dales JP, Garcia S, Meunier-Carpentier S, Andrac-Meyer L, Haddad O, Lavaut MN, Allasia C,
Bonnier P, Charpin C (2005) Overexpression of hypoxia-inducible factor HIF-1alpha predicts
early relapse in breast cancer: retrospective study in a series of 745 patients. Int J Cancer
116:734–739

Erler JT, Bennewith KL, Nicolau M, Dornhofer N, Kong C, Le QT, Chi JT, Jeffrey SS, Giaccia AJ
(2006) Lysyl oxidase is essential for hypoxia-induced metastasis. Nature 440:1222–1226

Forsythe JA, Jiang BH, Iyer NV, Agani F, Leung SW, Koos RD, Semenza GL (1996) Activation of
vascular endothelial growth factor gene transcription by hypoxia-inducible factor 1. Mol Cell
Biol 16:4604–4613

Fredlund E, Ringner M, Maris JM, Påhlman S (2008) High Myc pathway activity and low stage of
neuronal differentiation associate with poor outcome in neuroblastoma. Proc Natl Acad Sci
USA 105:14094–14099

Freeburg PB, Abrahamson DR (2003) Hypoxia-inducible factors and kidney vascular development. J Am Soc Nephrol 14:2723–2730

Galli R, Binda E, Orfanelli U, Cipelletti B, Gritti A, De Vitis S, Fiocco R, Foroni C, Dimeco F,
Vescovi A (2004) Isolation and characterization of tumorigenic, stem-like neural precursors
from human glioblastoma. Cancer Res 64:7011–7021

Generali D, Berruti A, Brizzi MP, Campo L, Bonardi S, Wigfield S, Bersiga A, Allevi G, Milani M,
Aguggini S, Gandolfi V, Dogliotti L, Bottini A, Harris AL, Fox SB (2006) Hypoxia-inducible
factor-1alpha expression predicts a poor response to primary chemoendocrine therapy and
disease-free survival in primary human breast cancer. Clin Cancer Res 12:4562–4568

Giatromanolaki A, Koukourakis MI, Sivridis E, Turley H, Talks K, Pezzella F, Gatter KC, Harris AL
(2001) Relation of hypoxia inducible factor 1 alpha and 2 alpha in operable non-small cell lung
cancer to angiogenic/molecular profile of tumours and survival. Br J Cancer 85:881–890

Giatromanolaki A, Sivridis E, Fiska A, Koukourakis MI (2006) Hypoxia-inducible factor-2 alpha
(HIF-2 alpha) induces angiogenesis in breast carcinomas. Appl Immunohistochem Mol
Morphol 14:78–82

Gnarra JR, Tory K, Weng Y, Schmidt L, Wei MH, Li H, Latif F, Liu S, Chen F, Duh FM et al
(1994) Mutations of the VHL tumour suppressor gene in renal carcinoma. Nat Genet 7:85–90

Groothuis DR, Pasternak JF, Fischer JM, Blasberg RG, Bigner DD, Vick NA (1983) Regional
measurements of blood flow in experimental RG-2 rat gliomas. Cancer Res 43:3362–3367

Gruber G, Greiner RH, Hlushchuk R, Aebersold DM, Altermatt HJ, Berclaz G, Djonov V (2004) Hypoxia-inducible factor 1 alpha in high-risk breast cancer: an independent prognostic parameter? Breast Cancer Res 6:R191–R198

Hansford LM, McKee AE, Zhang L, George RE, Gerstle JT, Thorner PS, Smith KM, Look AT, Yeger H, Miller FD, Irwin MS, Thiele CJ, Kaplan DR (2007) Neuroblastoma cells isolated from bone marrow metastases contain a naturally enriched tumor-initiating cell. Cancer Res 67:11234–11243

Heddleston JM, Li Z, McLendon RE, Hjelmeland AB, Rich JN (2009) The hypoxic microenvironment maintains glioblastoma stem cells and promotes reprogramming towards a cancer stem cell phenotype. Cell Cycle 8:3274–3284

Helczynska K, Kronblad A, Jögi A, Nilsson E, Beckman S, Landberg G, Påhlman S (2003) Hypoxia promotes a dedifferentiated phenotype in ductal breast carcinoma in situ. Cancer Res 63:1441–1444

Helczynska K, Larsson AM, Holmquist Mengelbier L, Bridges E, Fredlund E, Borgquist S, Landberg G, Påhlman S, Jirstrom K (2008) Hypoxia-inducible factor-2alpha correlates to distant recurrence and poor outcome in invasive breast cancer. Cancer Res 68:9212–9220

Hemmati HD, Nakano I, Lazareff JA, Masterman-Smith M, Geschwind DH, Bronner-Fraser M, Kornblum HI (2003) Cancerous stem cells can arise from pediatric brain tumors. Proc Natl Acad Sci USA 100:15178–15183

Henze AT, Riedel J, Diem T, Wenner J, Flamme I, Pouysseggur J, Plate KH, Acker T (2009) Prolyl hydroxylases 2 and 3 act in gliomas as protective negative feedback regulators of hypoxia-inducible factors. Cancer Res 70(1):357–366

Herman JG, Latif F, Weng Y, Lerman MI, Zbar B, Liu S, Samid D, Duan DS, Gnarra JR, Linehan WM et al (1994) Silencing of the VHL tumor-suppressor gene by DNA methylation in renal carcinoma. Proc Natl Acad Sci USA 91:9700–9704

Höckel M, Vaupel P (2001) Tumor hypoxia: definitions and current clinical, biologic, and molecular aspects. J Natl Cancer Inst 93:266–276

Hoehner JC, Gestblom C, Hedborg F, Sandstedt B, Olsen L, Påhlman S (1996) A developmental model of neuroblastoma: differentiating stroma-poor tumors' progress along an extra-adrenal chromaffin lineage. Lab Invest 75:659–675

Holmquist-Mengelbier L, Fredlund E, Löfstedt T, Noguera R, Navarro S, Nilsson H, Pietras A, Vallon-Christersson J, Borg A, Gradin K, Poellinger L, Påhlman S (2006) Recruitment of HIF-1alpha and HIF-2alpha to common target genes is differentially regulated in neuroblastoma: HIF-2alpha promotes an aggressive phenotype. Cancer Cell 10:413–423

Hossman KA, Bloink M (1981) Blood flow and regulation of blood flow in experimental peritumoral edema. Stroke 12:211–217

Hu CJ, Wang LY, Chodosh LA, Keith B, Simon MC (2003) Differential roles of hypoxia-inducible factor 1alpha (HIF-1alpha) and HIF-2alpha in hypoxic gene regulation. Mol Cell Biol 23:9361–9374

Huncharek M, Muscat J, Geschwind JF (1999) K-ras oncogene mutation as a prognostic marker in non-small cell lung cancer: a combined analysis of 881 cases. Carcinogenesis 20:1507–1510

Iyer NV, Kotch LE, Agani F, Leung SW, Laughner E, Wenger RH, Gassmann M, Gearhart JD, Lawler AM, Yu AY, Semenza GL (1998) Cellular and developmental control of O2 homeostasis by hypoxia-inducible factor 1 alpha. Genes Dev 12:149–162

Jensen RL (2006) Hypoxia in the tumorigenesis of gliomas and as a potential target for therapeutic measures. Neurosurg Focus 20:E24

Jögi A, Ora I, Nilsson H, Lindeheim A, Makino Y, Poellinger L, Axelson H, Påhlman S (2002) Hypoxia alters gene expression in human neuroblastoma cells toward an immature and neural crest-like phenotype. Proc Natl Acad Sci USA 99:7021–7026

Kaelin WG Jr (2002) Molecular basis of the VHL hereditary cancer syndrome. Nat Rev Cancer 2:673–682

Kaelin WG Jr, Ratcliffe PJ (2008) Oxygen sensing by metazoans: the central role of the HIF hydroxylase pathway. Mol Cell 30:393–402

Keith B, Simon MC (2007) Hypoxia-inducible factors, stem cells, and cancer. Cell 129:465–472

Kim SJ, Rabbani ZN, Dewhirst MW, Vujaskovic Z, Vollmer RT, Schreiber EG, Oosterwijk E, Kelley MJ (2005) Expression of HIF-1alpha, CA IX, VEGF, and MMP-9 in surgically resected non-small cell lung cancer. Lung Cancer 49:325–335

Kim MY, Oskarsson T, Acharyya S, Nguyen DX, Zhang XH, Norton L, Massague J (2009a) Tumor self-seeding by circulating cancer cells. Cell 139:1315–1326

Kim WY, Perera S, Zhou B, Carretero J, Yeh JJ, Heathcote SA, Jackson AL, Nikolinakos P, Ospina B, Naumov G, Brandstetter KA, Weigman VJ, Zaghlul S, Hayes DN, Padera RF, Heymach JV, Kung AL, Sharpless NE, Kaelin WG Jr, Wong KK (2009b) HIF2alpha cooperates with RAS to promote lung tumorigenesis in mice. J Clin Invest 119:2160–2170

Kondo K, Kim WY, Lechpammer M, Kaelin WG Jr (2003) Inhibition of HIF2alpha is sufficient to suppress pVHL-defective tumor growth. PLoS Biol 1:E83

Krieg M, Haas R, Brauch H, Acker T, Flamme I, Plate KH (2000) Up-regulation of hypoxia-inducible factors HIF-1alpha and HIF-2alpha under normoxic conditions in renal carcinoma cells by von Hippel-Lindau tumor suppressor gene loss of function. Oncogene 19:5435–5443

Kronblad A, Jirstrom K, Ryden L, Nordenskjold B, Landberg G (2006) Hypoxia inducible factor-1alpha is a prognostic marker in premenopausal patients with intermediate to highly differentiated breast cancer but not a predictive marker for tamoxifen response. Int J Cancer 118:2609–2616

Le QT, Chen E, Salim A, Cao H, Kong CS, Whyte R, Donington J, Cannon W, Wakelee H, Tibshirani R, Mitchell JD, Richardson D, O'Byrne KJ, Koong AC, Giaccia AJ (2006) An evaluation of tumor oxygenation and gene expression in patients with early stage non-small cell lung cancers. Clin Cancer Res 12:1507–1514

Leek RD, Talks KL, Pezzella F, Turley H, Campo L, Brown NS, Bicknell R, Taylor M, Gatter KC, Harris AL (2002) Relation of hypoxia-inducible factor-2 alpha (HIF-2 alpha) expression in tumor-infiltrative macrophages to tumor angiogenesis and the oxidative thymidine phosphorylase pathway in Human breast cancer. Cancer Res 62:1326–1329

Li Z, Bao S, Wu Q, Wang H, Eyler C, Sathornsumetee S, Shi Q, Cao Y, Lathia J, McLendon RE, Hjelmeland AB, Rich JN (2009) Hypoxia-inducible factors regulate tumorigenic capacity of glioma stem cells. Cancer Cell 15:501–513

Lidgren A, Hedberg Y, Grankvist K, Rasmuson T, Vasko J, Ljungberg B (2005) The expression of hypoxia-inducible factor 1alpha is a favorable independent prognostic factor in renal cell carcinoma. Clin Cancer Res 11:1129–1135

Löfstedt T, Fredlund E, Holmquist-Mengelbier L, Pietras A, Ovenberger M, Poellinger L, Påhlman S (2007) Hypoxia inducible factor-2alpha in cancer. Cell Cycle 6:919–926

Makino Y, Cao R, Svensson K, Bertilsson G, Asman M, Tanaka H, Cao Y, Berkenstam A, Poellinger L (2001) Inhibitory PAS domain protein is a negative regulator of hypoxia-inducible gene expression. Nature 414:550–554

Mandriota SJ, Turner KJ, Davies DR, Murray PG, Morgan NV, Sowter HM, Wykoff CC, Maher ER, Harris AL, Ratcliffe PJ, Maxwell PH (2002) HIF activation identifies early lesions in VHL kidneys: evidence for site-specific tumor suppressor function in the nephron. Cancer Cell 1:459–468

Matthay KK, Villablanca JG, Seeger RC, Stram DO, Harris RE, Ramsay NK, Swift P, Shimada H, Black CT, Brodeur GM, Gerbing RB, Reynolds CP (1999) Treatment of high-risk neuroblastoma with intensive chemotherapy, radiotherapy, autologous bone marrow transplantation, and 13-cis-retinoic acid. Children's Cancer Group. N Engl J Med 341:1165–1173

Maxwell PH, Wiesener MS, Chang GW, Clifford SC, Vaux EC, Cockman ME, Wykoff CC, Pugh CW, Maher ER, Ratcliffe PJ (1999) The tumour suppressor protein VHL targets hypoxia-inducible factors for oxygen-dependent proteolysis. Nature 399:271–275

Maynard MA, Evans AJ, Shi W, Kim WY, Liu FF, Ohh M (2007) Dominant-negative HIF-3 alpha 4 suppresses VHL-null renal cell carcinoma progression. Cell Cycle 6:2810–2816

McCord AM, Jamal M, Shankavaram UT, Lang FF, Camphausen K, Tofilon PJ (2009) Physiologic oxygen concentration enhances the stem-like properties of CD133+ human glioblastoma cells in vitro. Mol Cancer Res 7:489–497

Nilsson H, Jögi A, Beckman S, Harris AL, Poellinger L, Påhlman S (2005) HIF-2alpha expression in human fetal paraganglia and neuroblastoma: relation to sympathetic differentiation, glucose deficiency, and hypoxia. Exp Cell Res 303:447–456

Noguera R, Fredlund E, Piqueras M, Pietras A, Beckman S, Navarro S, Påhlman S (2009) HIF-1 alpha and HIF-2alpha are differentially regulated in vivo in neuroblastoma: high HIF-1alpha correlates negatively to advanced clinical stage and tumor vascularization. Clin Cancer Res 15:7130–7136

Ogden AT, Waziri AE, Lochhead RA, Fusco D, Lopez K, Ellis JA, Kang J, Assanah M, McKhann GM, Sisti MB, McCormick PC, Canoll P, Bruce JN (2008) Identification of A2B5 +CD133- tumor-initiating cells in adult human gliomas. Neurosurgery 62:505–514; discussion 514–515

Park SK, Dadak AM, Haase VH, Fontana L, Giaccia AJ, Johnson RS (2003) Hypoxia-induced gene expression occurs solely through the action of hypoxia-inducible factor 1alpha (HIF-1 alpha): role of cytoplasmic trapping of HIF-2alpha. Mol Cell Biol 23:4959–4971

Peng J, Zhang L, Drysdale L, Fong GH (2000) The transcription factor EPAS-1/hypoxia-inducible factor 2alpha plays an important role in vascular remodeling. Proc Natl Acad Sci USA 97:8386–8391

Percy MJ, Beer PA, Campbell G, Dekker AW, Green AR, Oscier D, Rainey MG, van Wijk R, Wood M, Lappin TR, McMullin MF, Lee FS (2008a) Novel exon 12 mutations in the HIF2A gene associated with erythrocytosis. Blood 111:5400–5402

Percy MJ, Furlow PW, Lucas GS, Li X, Lappin TR, McMullin MF, Lee FS (2008b) A gain-of-function mutation in the HIF2A gene in familial erythrocytosis. N Engl J Med 358:162–168

Pietras A, Gisselsson D, Ora I, Noguera R, Beckman S, Navarro S, Påhlman S (2008) High levels of HIF-2alpha highlight an immature neural crest-like neuroblastoma cell cohort located in a perivascular niche. J Pathol 214:482–488

Pietras A, Hansford LM, Johnsson AS, Bridges E, Sjolund J, Gisselsson D, Rehn M, Beckman S, Noguera R, Navarro S, Cammenga J, Fredlund E, Kaplan DR, Påhlman S (2009) HIF-2alpha maintains an undifferentiated state in neural crest-like human neuroblastoma tumor-initiating cells. Proc Natl Acad Sci USA 106:16805–16810

Rankin EB, Tomaszewski JE, Haase VH (2006) Renal cyst development in mice with conditional inactivation of the von Hippel-Lindau tumor suppressor. Cancer Res 66:2576–2583

Rankin EB, Biju MP, Liu Q, Unger TL, Rha J, Johnson RS, Simon MC, Keith B, Haase VH (2007) Hypoxia-inducible factor-2 (HIF-2) regulates hepatic erythropoietin in vivo. J Clin Invest 117:1068–1077

Rankin EB, Rha J, Unger TL, Wu CH, Shutt HP, Johnson RS, Simon MC, Keith B, Haase VH (2008) Hypoxia-inducible factor-2 regulates vascular tumorigenesis in mice. Oncogene 27:5354–5358

Raval RR, Lau KW, Tran MG, Sowter HM, Mandriota SJ, Li JL, Pugh CW, Maxwell PH, Harris AL, Ratcliffe PJ (2005) Contrasting properties of hypoxia-inducible factor 1 (HIF-1) and HIF-2 in von Hippel-Lindau-associated renal cell carcinoma. Mol Cell Biol 25:5675–5686

Raza SM, Lang FF, Aggarwal BB, Fuller GN, Wildrick DM, Sawaya R (2002) Necrosis and glioblastoma: a friend or a foe? A review and a hypothesis. Neurosurgery 51:2–12; discussion 12–13

Rosenberger C, Griethe W, Gruber G, Wiesener M, Frei U, Bachmann S, Eckardt KU (2003) Cellular responses to hypoxia after renal segmental infarction. Kidney Int 64:874–886

Ryan HE, Lo J, Johnson RS (1998) HIF-1 alpha is required for solid tumor formation and embryonic vascularization. EMBO J 17:3005–3015

Schindl M, Schoppmann SF, Samonigg H, Hausmaninger H, Kwasny W, Gnant M, Jakesz R, Kubista E, Birner P, Oberhuber G (2002) Overexpression of hypoxia-inducible factor 1alpha is associated with an unfavorable prognosis in lymph node-positive breast cancer. Clin Cancer Res 8:1831–1837

Scortegagna M, Morris MA, Oktay Y, Bennett M, Garcia JA (2003) The HIF family member EPAS1/HIF-2alpha is required for normal hematopoiesis in mice. Blood 102:1634–1640

Scortegagna M, Ding K, Zhang Q, Oktay Y, Bennett MJ, Bennett M, Shelton JM, Richardson JA, Moe O, Garcia JA (2005) HIF-2alpha regulates murine hematopoietic development in an erythropoietin-dependent manner. Blood 105:3133–3140

Semenza GL (2003) Targeting HIF-1 for cancer therapy. Nat Rev Cancer 3:721–732

Shweiki D, Itin A, Soffer D, Keshet E (1992) Vascular endothelial growth factor induced by hypoxia may mediate hypoxia-initiated angiogenesis. Nature 359:843–845

Singh SK, Clarke ID, Terasaki M, Bonn VE, Hawkins C, Squire J, Dirks PB (2003) Identification of a cancer stem cell in human brain tumors. Cancer Res 63:5821–5828

Singh SK, Hawkins C, Clarke ID, Squire JA, Bayani J, Hide T, Henkelman RM, Cusimano MD, Dirks PB (2004) Identification of human brain tumour initiating cells. Nature 432:396–401

Sowter HM, Raval RR, Moore JW, Ratcliffe PJ, Harris AL (2003) Predominant role of hypoxia-inducible transcription factor (Hif)-1alpha versus Hif-2alpha in regulation of the transcriptional response to hypoxia. Cancer Res 63:6130–6134

Swinson DE, Jones JL, Richardson D, Wykoff C, Turley H, Pastorek J, Taub N, Harris AL, O'Byrne KJ (2003) Carbonic anhydrase IX expression, a novel surrogate marker of tumor hypoxia, is associated with a poor prognosis in non-small-cell lung cancer. J Clin Oncol 21:473–482

Tan EY, Campo L, Han C, Turley H, Pezzella F, Gatter KC, Harris AL, Fox SB (2007) BNIP3 as a progression marker in primary human breast cancer; opposing functions in in situ versus invasive cancer. Clin Cancer Res 13:467–474

Tian H, McKnight SL, Russell DW (1997) Endothelial PAS domain protein 1 (EPAS1), a transcription factor selectively expressed in endothelial cells. Genes Dev 11:72–82

Tian H, Hammer RE, Matsumoto AM, Russell DW, McKnight SL (1998) The hypoxia-responsive transcription factor EPAS1 is essential for catecholamine homeostasis and protection against heart failure during embryonic development. Genes Dev 12:3320–3324

Volm M, Koomagi R (2000) Hypoxia-inducible factor (HIF-1) and its relationship to apoptosis and proliferation in lung cancer. Anticancer Res 20:1527–1533

Wiesener MS, Turley H, Allen WE, Willam C, Eckardt KU, Talks KL, Wood SM, Gatter KC, Harris AL, Pugh CW, Ratcliffe PJ, Maxwell PH (1998) Induction of endothelial PAS domain protein-1 by hypoxia: characterization and comparison with hypoxia-inducible factor-1alpha. Blood 92:2260–2268

Yohena T, Yoshino I, Takenaka T, Kameyama T, Ohba T, Kuniyoshi Y, Maehara Y (2009) Upregulation of hypoxia-inducible factor-1alpha mRNA and its clinical significance in non-small cell lung cancer. J Thorac Oncol 4:284–290

Zagzag D, Zhong H, Scalzitti JM, Laughner E, Simons JW, Semenza GL (2000) Expression of hypoxia-inducible factor 1alpha in brain tumors: association with angiogenesis, invasion, and progression. Cancer 88:2606–2618

Zimmer M, Doucette D, Siddiqui N, Iliopoulos O (2004) Inhibition of hypoxia-inducible factor is sufficient for growth suppression of VHL-/- tumors. Mol Cancer Res 2:89–95

Hypoxia and Hypoxia Inducible Factors in Cancer Stem Cell Maintenance

Zhizhong Li and Jeremy N. Rich

Contents

Abstract Hypoxia promotes tumor progression through multiple mechanisms including modifying angiogenesis, metabolism switch and invasion. Hypoxia inducible factors (HIFs), particularly HIF1α and HIF2α, are key mediators in cancer hypoxia response and high expression levels of HIFs correlate with a poor prognosis in various tumor types. Cancer stem cells (CSCs), also known as cancer initiating cells or tumor propagation cells, are neoplastic cells that could self-renewal, differentiate as well as initiate tumor growth in vivo. Cancer stem cells are believed to be the key drivers in tumor growth and therapy resistance. Hypoxia has been shown to help maintain multiple normal stem cell population but its roles in cancer stem cells were largely unknown. Our group and other researchers recently identified that hypoxia is also a critical microenvironmental factor in regulating cancer stem cells' self-renewal, partially by enhancing the activity of stem cell factors like Oct4, c-Myc and Nanog. The effects of hypoxia on cancer stem cells seem to be primarily mediated by HIFs, particularly HIF2α. HIF2α is highly expressed in cancer stem cells in gliomas and neuroblastomas and loss of

Z. Li
Department of Radiation Oncology, Duke University Medical Center, Durham, NC 27710, USA
J.N. Rich (✉)
Department of Stem Cell Biology and Regenerative Medicine, Lerner Research Institute, Cleveland Clinic, Cleveland, OH 44195, USA
e-mail: richj@ccf.org

M. Celeste Simon (ed.), *Diverse Effects of Hypoxia on Tumor Progression*,
Current Topics in Microbiology and Immunology 345, DOI 10.1007/82_2010_75
© Springer-Verlag Berlin Heidelberg 2010, published online: 26 June 2010

HIF2α leads to significant decrease in cancer stem cell proliferation and self-renewal. These findings illustrated a new mechanism through which oxygen tension and microenvironment influences cancer development. Targeting hypoxia niches may therefore improve therapy efficacy by eliminating cancer stem cell population.

1 Hypoxia in Tumor Progression

Hypoxia, defined as reduced oxygen tension, is a common physiological phenomenon in both normal embryonic development and malignancy progression (Harris 2002; Bertout et al. 2008). Although severe hypoxia is generally toxic for both normal tissue and tumors, neoplastic cells gradually adapt to prolonged hypoxia though additional genetic and genomic changes with a net result that hypoxia promotes tumor progression and therapeutic resistance. Markers of hypoxia informs poor prognosis in numerous tumor types (Harris 2002; Kaur et al. 2005; Pouyssegur et al. 2006; Bertout et al. 2008). Using electrodes to measure the oxygen tension within solid tumors, Vaupel and colleagues reported that low oxygen levels were associated with increased tumor growth and risk of metastasis with greater risk of adverse outcomes in patients suffering from head and neck, cervical and breast cancers (Hockel and Vaupel 2001; Vaupel and Mayer 2007). Hypoxia has been extensively studied in other cancers including brain tumors. In glioblastoma multiforme, necrosis, a well known consequence of prolonged hypoxia, is one of the hallmarks of this highly lethal disease.

Hypoxia promotes cancer progression by regulating various aspects of cancer biology, including radiotherapy resistance, metabolism, angiogenesis and invasion/migration (Harris 2002). A major focus of hypoxia research in early era was its role in radiotherapy responses. As early as 1909, Schwarz and colleagues already noted that hypoxic cells were more resistant to ionic radiation than those irradiated in the presence of O_2 (Schwarz 1909). Hypoxia also significantly influences many other crucial steps in cancer progression. Under hypoxia conditions, cells switch their glucose metabolism from the aerobic tricarboxylic acid (TCA) cycle to anaerobic glycolysis. Glycolysis fuels tumor cell growth because glycolytic pathway provide the precursors for synthesis of nucleotides and phospholipids, both of which are essential for rapid cell growth (Bertout et al. 2008). In fact, cancer cells display a preference for glycolytic metabolism even in the presence of oxygen, a phenomenon commonly known as Warburg effect (Warburg 1956). Another well characterized biological consequence of tumor hypoxia is elevated angiogenesis. Upon low oxygen stimulation, tumor cells stimulate endothelial cell proliferation/migration and new blood vessel growth by secreting remarkable level of many pro-angiogenic growth factors, such as vascular endothelial growth factor (VEGF), platelet derived growth factor (PDGF), etc. In addition, hypoxia may also change the level of cell adhesion molecules and proteinases (like matrix metalloproteinases) to facilitate cancer invasion and migration (Harris 2002). Thus, hypoxia has broad influence on tumor biology and its new roles in the malignant progression are still under active investigation.

2   Hypoxia Inducible Factors and Tumor Progression

Cellular responses to oxygen tension are complicated processes that involve large molecular networks, including the hypoxia inducible factors (HIFs). The HIFs, heterodimer molecules consisting of an alpha subunit and a beta subunit, which are believed to exert pivotal roles in hypoxic responses (Kaur et al. 2005). HIFα protein levels are tightly regulated by oxygen while HIFβ is constitutively expressed in most cell types regardless of oxygen level (Jaakkola et al. 2001). Under "normoxia" (often mistakenly considered ambient oxygen levels, whereas physiologic oxygen levels are often 0.5–7%), HIFα is hydroxylated by PHD domain containing proteins, ubiquitinated by E3 ligase von Hippel-Lindau (VHL) and then degraded by then proteasome (Maxwell et al. 1999; Ivan et al. 2001; Jaakkola et al. 2001). In humans and mice, there are at least three HIFα subunits – HIF1α, HIF2α and HIF3α. HIF1α and HIF2α usually function as transcriptional activators with hundreds of downstream targets identified as regulated by HIF1α and/or HIF2α (Wang et al. 2005; Carroll and Ashcroft 2006; Pouyssegur et al. 2006; Bertout et al. 2008). HIF3α shares a DNA binding domain with HIF1α and HIF2α but lacks the transcription activation domain. Therefore, HIF3α is considered to largely serve a dominant negative role in hypoxia response by preventing the binding of HIF1α and/or HIF2α to their target promoters (Makino et al. 2001).

While HIF1α and HIF2α share significant homology, most studies have been focused on HIF1α largely due to its earlier discovery and more universal expression pattern in tissues than HIF2α (Ema et al. 1997). HIF1α functions as an oncogene (with notable exceptions) in many types of solid tumors, including those in breast, colon and brain. HIF1α promotes tumor angiogenesis, proliferation, glycolysis, and metastasis (Bertout et al. 2008; Yang et al. 2008). HIF1α protein levels are informative negative prognostic factors for many cancers and high levels of HIF1α correlate with worse patient outcome (Vaupel and Mayer 2007; Hoffmann et al. 2008). In contrast, the role of HIF2α in tumorigenesis has been less studied and poorly defined. Earlier data supported HIF2α as a tumor suppressor in neuroblastoma despite the promotion of angiogenesis (Acker et al. 2005); however, recent research using a von Hippel-Lindau (VHL)-deficient renal cancer model suggests that HIF2α promotes tumor proliferation and radiation resistance (Gordan et al. 2007; Bertout et al. 2009). Moreover, like HIF1α, high HIF2α expression correlates with worse prognosis in multiple cancer types including nonsmall cell lung cancer, breast cancer and hepatocellular carcinomas (Giatromanolaki et al. 2001; Bangoura et al. 2007; Helczynska et al. 2008). Therefore, HIF2α may function as an oncogene in certain contexts.

3   Hypoxia in Stem Cell and Cancer Stem Cell Maintenance

Two features of stem cells are self-renewal and differentiation. Normal stem cells use these two properties to sustain the tissue organization and hierarchy. Although tumors were traditionally thought to be highly disorganized, increasing evidence has accumulated that cellular hierarchies exist in many tumor types (Reya et al. 2001). Cancer stem cells, also known as cancer initiating cells or tumor propagation cells, are cellular subpopulations in cancers that share many characteristics with normal somatic stem/progenitor cells such as self-renewal capability, expression of stem cell markers, and multilineage differentiation (Reya et al. 2001; Bao et al. 2006a, b). The existence of cancer stem cells was first demonstrated in leukemia (Bonnet and Dick 1997); since then similar populations have been prospectively identified in multiple solid tumor types including breast cancer, brain tumors, and colon cancers (Al-Hajj et al. 2003; Singh et al. 2003; O'Brien et al. 2007). While the detailed mechanisms by which cancer stem cells promote tumorigenesis remain to be elucidated, our lab has shown that glioma stem cell can promote tumor angiogenesis and therapy resistance (Bao et al. 2006a, b).

Stem cells are maintained in special microenvironments termed niches (Zhang and Li 2008). The functional components of a stem cell niche are still under investigation but vasculature and endothelial cells seem to be critical for the maintenance of stem cells in adipose tissue, the nervous system, bone marrow and testes (Yoshida et al. 2007; Kiel and Morrison 2008; Tang et al. 2008; Tavazoie et al. 2008). Like their normal counterparts, cancer stem cells are also believed to rely on their own niches to sustain the population (Gilbertson and Rich 2007). It has been suggested that disruption of tumor microenvironment may serve as a critical therapeutic paradigm to kill tumor cells. Therefore, how cancer stem cells are maintained *in vivo* and how to destroy cancer stem cell niches are important questions for both basic research and drug development.

Oxygen tension is an important component of tissue microenvironment and local oxygen concentrations can directly influence stem cell self-renewal and differentiation. *In vitro* evidence indicates that hypoxia promotes an undifferentiated status. Low oxygen maintains embryonic stem cells and significantly blocks spontaneous cell differentiation (Ezashi et al. 2005). Culturing human hematopoietic stem cells (HSCs) under hypoxic conditions promotes their ability to repopulate when they are transplanted into nonobese diabetic (NOD)/severe combined immunodeficiency (SCID) mice (Danet et al. 2003). In fact, the bone marrow, where HSCs usually locate, is generally hypoxic (Chow et al. 2001). Hypoxia also regulates neural stem cells (NSCs) as low oxygen promotes the proliferation and survival of NSCs. Hypoxia alters the differentiation program of NSC to favor the generation of dopaminergic neurons (Studer et al. 2000). Why hypoxia stimulates the maintenance of stem cells is largely unknown but one attractive hypothesis is that stem cells are located in low oxygen environment to reduce the DNA damage resulting from reactive oxygen species (ROS). Recent reports have also identified a few molecular mechanisms by which hypoxia and HIFs directly modify stem cell function. For instance, Notch signaling is essential in hypoxia-mediated differentiation blockade in myogenic satellite cells and primary NSCs (Gustafsson et al. 2005). Simon and colleagues reported that HIF2α but not HIF1α up-regulates the expression of Oct4 and enhances the activity of c-myc (Covello et al. 2006; Gordan et al. 2007). As both Oct4 and c-myc are factors modulating stem cell self-renewal, these data shed light on how hypoxia helps maintain stem cells. Consistent with these findings, hypoxia augments the efficiency of forming induced pluripotent stem cells (iPSCs) (Yoshida et al. 2009).

Interestingly, hypoxia also directly regulates cancer stem cells and affects cancer progression. Hypoxia promotes an immature phenotype in solid tumors including human neuroblastoma and breast cancer cells (Axelson et al. 2005). Recently, our lab and others demonstrates that hypoxia directly helps maintain the cancer stem cell populations in brain tumors. We found that hypoxia promotes the self-renewal capability of glioma stem cells. Cancer stem cells cultured in ≤2% oxygen concentration display less spontaneous differentiation than those in 21% oxygen. Low oxygen tension may convert nonstem cancer cells into cancer stem cell-like status with increased self-renewal capacity as well as induction of essential stem cell factors, such as Oct4, Nanog, and c-Myc (Heddleston et al. 2009). Similarly, hypoxia promotes expansion of the glioma stem cells by enhancing their self-renewal while inhibiting differentiation (Soeda et al. 2009). In addition, colon cancer stem cells also preferentially locate in hypoxia niches (Zhang et al. 2008). We found that HIF1α is expressed in both cancer stem cell and nonstem cancer cells upon induction of hypoxia. In contrast, HIF2α is only highly induced in cancer stem cell populations. Both HIF1α and HIF2α are critical for cancer stem cell maintenance, as knockdown of either HIF1α or HIF2α in cancer stem cells leads to reduced self-renewal capacity, increased apoptosis and attenuated tumorigenesis (Li et al. 2009). Park and colleagues showed that HIF1α knockdown disrupted the effects of hypoxia on enhancing the *in vitro* self-renewal of glioma cancer stem cells and the inhibition of differentiation (Soeda et al. 2009). The unique expression pattern of HIF2α in cancer stem cells made us hypothesize that it plays central roles in cancer stem cell maintenance. Indeed, overexpression of nondegradable HIF2α protein in nonstem cancer cells, which normally do not express HIF2α protein, promotes a cancer stem cell-like phenotype and significantly increased their tumorigenic potential (Heddleston et al. 2009). In agreement with our data, Pahlman and colleagues recently reported that HIF2α maintains an undifferentiated state of neuroblastoma tumor initiating cells. Knockdown of HIF2α in neuroblastoma samples impaired tumorigenesis and led to a more differentiated and less aggressive tumor phenotype (Pietras et al. 2009). Interestingly, HIF2α is not expressed in normal NSCs (Li et al. 2009), making it a very attractive target for drug development. The relative importance of HIF1α and HIF2α in cancer stem cell maintenance remains unclear.

4 Hypoxia Niche Versus Vascular Niche: Three Different Models

Increasing evidence demonstrates that vasculature is a functional component of niches for both normal stem cell and cancer stem cells. Endothelial cells promote NSC self-renewal and maintenance (Shen et al. 2008). Adipocyte progenitors and undifferentiated spermatogonia are also shown to preferentially locate adjacent to blood vessels (Yoshida et al. 2007; Tang et al. 2008). Interestingly, this dependence on vascular structure based niche seems to be preserved in cancer development. Cancer stem cells from several types of brain tumors (medulloblastoma, ependymoma, oligodendroglioma and glioblastoma) seem to be associated with blood vessels and disturbing this vascular niche significantly inhibits the self-renewal of cancer stem cells *in vitro* and their tumorigenesis capability *in vivo* (Calabrese et al. 2007; Gilbertson and Rich 2007). Recently, our lab and others have begun to uncover the important role of hypoxia and hypoxia inducible factor in maintaining cancer stem cells (Li et al. 2009). At first glance, it seems challenging to consolidate these two facts that both blood vessels and hypoxic environment/pathways maintain cancer stem cells. There are at least three reasonable hypotheses for this dilemma.

The first and easiest explanation would be there are actually two distinct niches for cancer stem cells: a vascular niche and a hypoxic niche. It has been shown that there are actually two distinct niches for hematopoietic stem cells in bone marrow: an "osteoblastic" niche and a "vascular" niche (Kiel and Morrison 2008). It has also been proposed that HSC populations maintained by these two niches are different. The "osteoblastic" niche support long-term HSC maintenance and the "vascular" niche stimulates short-term HSC proliferation and balances multiple-lineage fate specification (Perry and Li 2007). It will be interesting to investigate whether it is true in the cancer stem cell scenario. Several related questions remain open: whether there are distinct cancer stem cell populations in a given tumor? Whether different cancer stem cell populations preferentially rely on one niche over the other?

Secondly, it is possible that the vascular and hypoxic niches are integrated. While common sense assumes that regions adjacent to blood vessels are generally not hypoxic, it is more complicated in tumors. Dewhirst and colleagues have shown that tumor-associated blood vessels can also be hypoxic as a result of many factors such as insufficient arteriolar supply, dysregulated microvessels and extreme fluctuation of red blood cell flux, etc. (Kimura et al. 1996). Therefore, it is plausible to hypothesize that cancer stem cells may choose to locate close to hypoxic blood vessels where they could remain under low oxygen tension to reduce ROS-induced cell stress and damage while at the same time, receive growth factors from vasculature (e.g., endothelial cells and/or pericytes) to maintain their undifferentiated status. To prove this hypothesis will require technical advances, including sensitive methods to mark hypoxic gradients in tumors and robust markers for cancer stem cells.

Third, it is possible that HIFs maintain cancer stem cells in normoxia conditions. As suggested by their name, HIF1α and HIF2α are usually induced by low oxygen tension. However, HIFs, particularly HIF2α can be activated even in the absence of hypoxia (Holmquist-Mengelbier et al. 2006; Li et al. 2009). *In vivo* hypoxia is generally defined as oxygen levels below 1%, in which both HIF1α and HIF2α proteins are stabilized. However, our group and others have demonstrated that HIF2α but not HIF1α accumulate even when oxygen concentrations are up to 5%. Because 5% oxygen is generally considered as "physiological normoxia" *in vivo*, it suggests that unlike HIF1α, HIF2α functional in the tissues under nonhypoxic conditions. HIF2α may then help maintain cancer stem cells even if they locate in well-oxygenated regions (e.g., adjacent to functional blood vessels). How HIF2α is induced under normoxia is poorly understood but it has been shown that oncogenes such as mitogen-activated protein kinase (MAPK) family members activate HIF2α by protein phosphorylation (Conrad et al. 1999). If oncogenes turn on HIF2α in cancer stem cells, HIF2α could in turn stimulate the secretion of pro-angiogenic factors such as VEGF and therefore induce new blood vessel growth. As a result, cancer stem cells create their own vascular niche but remain dependent on the niche (Gilbertson and Rich 2007). As long as the oxygen level is below 5% in cells with oncogenic growth factor pathway activation, HIF2α expression may be maintained and contribute to cancer stem cell growth.

5   Conclusion

The identification of cancer stem cells significantly nuances our views on cancer development and progression. Given the strong capability of cancer stem cell to initiate tumorigenesis and promote therapeutic resistance, it is critical for researchers to identify how this population is maintained *in vivo*. Cancer stem cells and normal stem cells share many properties, including the requirement of specific niches for their function. Disturbance of the stem cell/cancer stem cell niche could lead to their differentiation and loss of stemness characteristics (Gilbertson and Rich 2007). In cancer stem cell scenario, this could lead to decreased tumor growth and improved response to therapy. Many signaling molecules that mediate the communication between the microenvironment and normal stem cells are also indispensable for cancer stem cell maintenance. The best studied examples include sonic hedgehog, wnt and Notch pathways (Clement et al. 2007; Malanchi et al. 2008; Bolos et al. 2009). Hypoxia promotes normal stem cell maintenance, but the roles of hypoxia and HIFs in cancer stem cell maintenance have just begun to be appreciated. The discovery of the importance of hypoxia and HIFs in cancer stem cells has significant implications. On one hand, it indicates that the current cell culture conditions for cancer cells using ambient air conditions are not ideal and may create a loss of cellular diversity that was present in the parental tumor. Ambient air conditions submit cells to hyperoxic stress, which may lead to irreversible genetic or epigenetic changes that do not represent *in vivo* situation. To study cancer heterogeneity, *in vitro* assays should be conducted in restricted oxygen conditions. Stem cell biologists realize this and we believe this concept should be emphasized in cancer research as well (Wion et al. 2009). On the other hand, the importance of hypoxia, particularly HIFs in cancer stem cell provides us new targets for anti-cancer stem cell drug discovery. In malignant gliomas, HIF2α seems one of the better targets as it is highly induced in cancer stem cells but absent in normal tissues including normal stem cells. Further studies are needed to clarify whether HIF2α is also expressed and indispensable in cancer stem cells of other tumor types and identify the best strategies to target this conventionally "undrugable" transcriptional factor.

References

Acker T, Diez-Juan A et al (2005) Genetic evidence for a tumor suppressor role of HIF-2alpha. Cancer Cell 8(2):131–141

Al-Hajj M, Wicha MS et al (2003) Prospective identification of tumorigenic breast cancer cells. Proc Natl Acad Sci USA 100(7):3983–3988

Axelson H, Fredlund E et al (2005) Hypoxia-induced dedifferentiation of tumor cells–a mechanism behind heterogeneity and aggressiveness of solid tumors. Semin Cell Dev Biol 16(4–5):554–563

Bangoura G, Liu ZS et al (2007) Prognostic significance of HIF-2alpha/EPAS1 expression in hepatocellular carcinoma. World J Gastroenterol 13(23):3176–3182

Bao S, Wu Q et al (2006a) Glioma stem cells promote radioresistance by preferential activation of the DNA damage response. Nature 444(7120):756–760

Bao S, Wu Q et al (2006b) Stem cell-like glioma cells promote tumor angiogenesis through vascular endothelial growth factor. Cancer Res 66(16):7843–7848

Bertout JA, Majmundar AJ et al (2009) HIF2alpha inhibition promotes p53 pathway activity, tumor cell death, and radiation responses. Proc Natl Acad Sci USA 106(34):14391–14396

Bertout JA, Patel SA et al (2008) The impact of O2 availability on human cancer. Nat Rev Cancer 8(12):967–975

Bolos V, Blanco M et al (2009) Notch signalling in cancer stem cells. Clin Transl Oncol 11(1):11–19

Bonnet D, Dick JE (1997) Human acute myeloid leukemia is organized as a hierarchy that originates from a primitive hematopoietic cell. Nat Med 3(7):730–737

Calabrese C, Poppleton H et al (2007) A perivascular niche for brain tumor stem cells. Cancer Cell 11(1):69–82

Carroll VA, Ashcroft M (2006) Role of hypoxia-inducible factor (HIF)-1alpha versus HIF-2alpha in the regulation of HIF target genes in response to hypoxia, insulin-like growth factor-I, or loss of von Hippel-Lindau function: implications for targeting the HIF pathway. Cancer Res 66 (12):6264–6270

Chow DC, Wenning LA et al (2001) Modeling pO(2) distributions in the bone marrow hemato-poietic compartment. II. Modified Kroghian models. Biophys J 81(2):685–696

Clement V, Sanchez P et al (2007) HEDGEHOG-GLI1 signaling regulates human glioma growth, cancer stem cell self-renewal, and tumorigenicity. Curr Biol 17(2):165–172

Conrad PW, Freeman TL et al (1999) EPAS1 trans-activation during hypoxia requires p42/p44 MAPK. J Biol Chem 274(47):33709–33713

Covello KL, Kehler J et al (2006) HIF-2alpha regulates Oct-4: effects of hypoxia on stem cell function, embryonic development, and tumor growth. Genes Dev 20(5):557–570

Danet GH, Pan Y et al (2003) Expansion of human SCID-repopulating cells under hypoxic conditions. J Clin Invest 112(1):126–135

Ema M, Taya S et al (1997) A novel bHLH-PAS factor with close sequence similarity to hypoxia-inducible factor 1alpha regulates the VEGF expression and is potentially involved in lung and vascular development. Proc Natl Acad Sci USA 94(9):4273–4278

Ezashi T, Das P et al (2005) Low O2 tensions and the prevention of differentiation of hES cells. Proc Natl Acad Sci USA 102(13):4783–4788

Giatromanolaki A, Koukourakis MI et al (2001) Relation of hypoxia inducible factor 1 alpha and 2 alpha in operable non-small cell lung cancer to angiogenic/molecular profile of tumours and survival. Br J Cancer 85(6):881–890

Gilbertson RJ, Rich JN (2007) Making a tumour's bed: glioblastoma stem cells and the vascular niche. Nat Rev Cancer 7(10):733–736

Gordan JD, Bertout JA et al (2007) HIF-2alpha promotes hypoxic cell proliferation by enhancing c-myc transcriptional activity. Cancer Cell 11(4):335–347

Gustafsson MV, Zheng X et al (2005) Hypoxia requires notch signaling to maintain the undifferentiated cell state. Dev Cell 9(5):617–628

Harris AL (2002) Hypoxia–a key regulatory factor in tumour growth. Nat Rev Cancer 2(1): 38–47

Heddleston JM, Li Z et al (2009) The hypoxic microenvironment maintains glioblastoma stem cells and promotes reprogramming towards a cancer stem cell phenotype. Cell Cycle 8 (20):3274–3284

Helczynska K, Larsson AM et al (2008) Hypoxia-inducible factor-2alpha correlates to distant recurrence and poor outcome in invasive breast cancer. Cancer Res 68(22):9212–9220

Hockel M, Vaupel P (2001) Tumor hypoxia: definitions and current clinical, biologic, and molecular aspects. J Natl Cancer Inst 93(4):266–276

Hoffmann AC, Mori R et al (2008) High expression of HIF1a is a predictor of clinical outcome in patients with pancreatic ductal adenocarcinomas and correlated to PDGFA, VEGF, and bFGF. Neoplasia 10(7):674–679

Holmquist-Mengelbier L, Fredlund E et al (2006) Recruitment of HIF-1alpha and HIF-2alpha to common target genes is differentially regulated in neuroblastoma: HIF-2alpha promotes an aggressive phenotype. Cancer Cell 10(5):413–423

Ivan M, Kondo K et al (2001) HIFalpha targeted for VHL-mediated destruction by proline hydroxylation: implications for O2 sensing. Science 292(5516):464–468

Jaakkola P, Mole DR et al (2001) Targeting of HIF-alpha to the von Hippel-Lindau ubiquitylation complex by O2-regulated prolyl hydroxylation. Science 292(5516):468–472

Kaur B, Khwaja FW et al (2005) Hypoxia and the hypoxia-inducible-factor pathway in glioma growth and angiogenesis. Neuro Oncol 7(2):134–153

Kiel MJ, Morrison SJ (2008) Uncertainty in the niches that maintain haematopoietic stem cells. Nat Rev Immunol 8(4):290–301

Kimura H, Braun RD et al (1996) Fluctuations in red cell flux in tumor microvessels can lead to transient hypoxia and reoxygenation in tumor parenchyma. Cancer Res 56(23):5522–5528

Li Z, Bao S et al (2009) Hypoxia-inducible factors regulate tumorigenic capacity of glioma stem cells. Cancer Cell 15(6):501–513

Makino Y, Cao R et al (2001) Inhibitory PAS domain protein is a negative regulator of hypoxia-inducible gene expression. Nature 414(6863):550–554

Malanchi I, Peinado H et al (2008) Cutaneous cancer stem cell maintenance is dependent on beta-catenin signalling. Nature 452(7187):650–653

Maxwell PH, Wiesener MS et al (1999) The tumour suppressor protein VHL targets hypoxia-inducible factors for oxygen-dependent proteolysis. Nature 399(6733):271–275

O'Brien CA, Pollett A et al (2007) A human colon cancer cell capable of initiating tumour growth in immunodeficient mice. Nature 445(7123):106–110

Perry JM, Li L (2007) Disrupting the stem cell niche: good seeds in bad soil. Cell 129(6):1045–1047

Pietras A, Hansford LM et al (2009) HIF-2alpha maintains an undifferentiated state in neural crest-like human neuroblastoma tumor-initiating cells. Proc Natl Acad Sci USA 106 (39):16805–16810

Pouyssegur J, Dayan F et al (2006) Hypoxia signalling in cancer and approaches to enforce tumour regression. Nature 441(7092):437–443

Reya T, Morrison SJ et al (2001) Stem cells, cancer, and cancer stem cells. Nature 414 (6859):105–111

Schwarz G (1909) Ueber Desensibilisierung gegen röntgen- und radiumstrahlen. Munchener Medizinische Wochenschrift 24:1–2

Shen Q, Wang Y et al (2008) Adult SVZ stem cells lie in a vascular niche: a quantitative analysis of niche cell-cell interactions. Cell Stem Cell 3(3):289–300

Singh SK, Clarke ID et al (2003) Identification of a cancer stem cell in human brain tumors. Cancer Res 63(18):5821–5828

Soeda A, Park M et al (2009) Hypoxia promotes expansion of the CD133-positive glioma stem cells through activation of HIF-1alpha. Oncogene 28(45):3949–3959

Studer L, Csete M et al (2000) Enhanced proliferation, survival, and dopaminergic differentiation of CNS precursors in lowered oxygen. J Neurosci 20(19):7377–7383

Tang W, Zeve D et al (2008) White fat progenitor cells reside in the adipose vasculature. Science 322(5901):583–586

Tavazoie M, Van der Veken L et al (2008) A specialized vascular niche for adult neural stem cells. Cell Stem Cell 3(3):279–288

Vaupel P, Mayer A (2007) Hypoxia in cancer: significance and impact on clinical outcome. Cancer Metastasis Rev 26(2):225–239

Wang V, Davis DA et al (2005) Differential gene up-regulation by hypoxia-inducible factor-1 alpha and hypoxia-inducible factor-2alpha in HEK293T cells. Cancer Res 65(8):3299–3306

Warburg O (1956) On the origin of cancer cells. Science 123(3191):309–314

Wion D, Christen T et al (2009) PO(2) matters in stem cell culture. Cell Stem Cell 5(3):242–243

Yang MH, Wu MZ et al (2008) Direct regulation of TWIST by HIF-1alpha promotes metastasis. Nat Cell Biol 10(3):295–305

Yoshida S, Sukeno M et al (2007) A vasculature-associated niche for undifferentiated spermatogonia in the mouse testis. Science 317(5845):1722–1726

Yoshida Y, Takahashi K et al (2009) Hypoxia enhances the generation of induced pluripotent stem cells. Cell Stem Cell 5(3):237–241

Zhang J, Li L (2008) Stem cell niche: microenvironment and beyond. J Biol Chem 283(15): 9499–9503

Zhang Y, Guo W et al (2008) Human colon cancer stem cells locate in hypoxic niche. J Clin Oncol, 2008 ASCO Annual meeting proceedings 26(15S):Abstract 22209

Role of Carcinoma-Associated Fibroblasts and Hypoxia in Tumor Progression

Amato J. Giaccia and Ernestina Schipani

Contents

Abstract In recent years, a variety of experimental evidence has convincingly shown that progression of malignant tumors does not depend exclusively on cell-autonomous properties of the cancer cells, but can also be influenced by the tumor stroma. The concept that cancer cells are subjected to microenvironmental control has thus emerged as an important chapter in cancer biology. Recent findings have suggested an important role, in particular, for macrophages, endothelial cells, and

A.J. Giaccia (✉)
Department of Radiation Oncology, Division of Cancer and Radiation Biology, Stanford University School of Medicine, Stanford, CA 94305-5152, USA
e-mail: giaccia@stanford.edu
E. Schipani
Department of Medicine, Endocrine Unit, MGH-Harvard Medical School, Boston, MA 02114, USA

M. Celeste Simon (ed.), *Diverse Effects of Hypoxia on Tumor Progression*,
Current Topics in Microbiology and Immunology 345, DOI 10.1007/82_2010_73
© Springer-Verlag Berlin Heidelberg 2010, published online: 2 June 2010

cancer-associated fibroblasts (CAFs) in tumor growth and progression. Numerous lines of evidence indicate that the bone marrow is the source, at least in part, of these cells. This chapter summarizes our current knowledge of how bone marrow contributes to the tumor stroma, with particular emphasis on CAFs. The potential role of hypoxia in modulating the differentiation and activity of CAFs, and the therapeutical implications of targeting CAFs for anticancer therapy are discussed.

1   Introduction

Untransformed cells require a specific environmental niche for their growth and survival. The concept that transformed cells are subjected to microenvironmental control and that the tumor microenvironment is a crucial component of cancer progression is now emerging (Hanahan and Weinberg 2000; Coussens and Werb 2002; De Wever and Mareel 2003; Tlsty and Coussens 2006; De Wever et al. 2008; Lorusso and Ruegg 2008). A link between inflammation and cancer was already recognized in 1863 by Rudolph Virchow, when he reported the presence of leucocytes in tumor tissue (Balkwill and Mantovani 2001). In recent years, a variety of experimental evidence has convincingly shown that progression of malignant tumors does not depend exclusively on cell-autonomous properties of the cancer cells, but can also be influenced by the tumor stroma, namely by the compartment that provides the connective-tissue framework of the tumor itself (Bissell et al. 2002; Coussens and Werb 2002).

The tumor stroma is formed by a variety of cells including fibroblasts, immune and inflammatory cells, and adipocytes, which are all embedded in an extracellular matrix (ECM) and nourished by an enriched vascular network. In its cellular make-up, it thus resembles the granulation tissue of healing wounds. Notably, malignant tumors have also been characterized as "wounds that never heal" (Dvorak 1986). However, the amount of stroma and its composition vary from tumor to tumor. In some cases of commonly occurring carcinomas, the nonneoplastic cells may account for as many as 90% of the cells within the tumor mass (Weinberg 2007).

The interactions between cancer cells and the surrounding stroma are complex, and only partially elucidated. Cancer cells alter their adjacent stroma by producing a large range of growth factors and proteases. These factors activate adjacent stromal cells in a paracrine manner, and thus cause the secretion by these cells of another array of growth factors and proteases, which in turn affect tumor growth and metastasis. This "vicious cycle" has become the target of novel therapeutic approaches to cancer (Liotta and Kohn 2001; Albini and Sporn 2007; Pietras et al. 2008).

Recent findings have suggested an important role, in particular, for macrophages, endothelial cells, and cancer-associated fibroblasts (CAFs) in tumor growth and progression. Numerous lines of evidence indicate that the bone marrow is the source, at least in part, of these cells. This chapter summarizes our current knowledge about the contribution of the bone marrow to the tumor stroma, with particular emphasis on carcinoma-associated fibroblasts (CAFs). The potential role of hypoxia in modulating the differentiation and activity of CAFs is discussed.

2 Adult Bone Marrow

2.1 Hematopoietic Stem Cells, Mesenchymal Stem Cells and Endothelial Progenitors Cells

The bone marrow comprises both hematopoietic and mesenchymal populations. Hematopoietic cells derive from self-renewing hematopoietic stem cells (HSCs), and constitute the vast majority of the adult bone marrow's cellularity, whereas the mesenchymal population is thought to originate from mesenchymal stem cells (MSCs) (Schipani and Kronenberg 2008). The presence of nonhematopoietic stem cells in the bone marrow was first suggested by the German pathologist Cohneim about 130 years ago. He proposed that bone marrow can be the source of fibroblasts contributing to wound healing in numerous peripheral tissues (Prockop 1997). In the early 1970s, the pioneering work of Friedenstein and colleagues demonstrated that rodent bone marrow had fibroblastoid cells with clonogenic potential *in vitro* (Friedenstein et al. 1970, 1980). Over the years, numerous laboratories have confirmed and expanded these findings by showing that cells isolated according to Friedenstein's protocol were also present in the human bone marrow and by demonstrating that these cells could be subpassaged and differentiated *in vitro* into a variety of cells of the mesenchymal lineages (Prockop 1997; Pittenger et al. 1999; Caplan 2007; Kolf et al. 2007; Bianco et al. 2008). Friedenstein had thus isolated from the bone marrow what later on would be renamed "mesenchymal stem cell" or MSC.

The current model proposes that there are at least two types of stem cells in the bone marrow: HSCs and MSCs. HSCs would give rise to hematopoietic cell types and to cells that resorb bone (osteoclasts), whereas MSCs would differentiate into a variety of mesenchymal lineages such as chondrocytes, adipocytes, and osteoblasts, at least *in vitro*. However, recent experimental evidence has indicated that MSCs are likely to be just a subset of a heterogeneous population of nonhematopoietic cells in the adult bone marrow. Endothelial progenitor cells (EPCs) constitute another subset of bone marrow nonhematopoietic cells. EPCs are considered endothelial precursors that reside in the bone marrow and are released into the bloodstream to contribute to vasculogenesis in injured organs (Asahara et al. 1999).

2.2 Fibrocyte Progenitor Cells

Another cell type of bone marrow origin and referred to as "fibrocyte" has been recently characterized (Bucala et al. 1994; Quan et al. 2004; Barth and Westhoff 2007; Bellini and Mattoli 2007; Quan and Bucala 2007). Fibrocytes are collagen-producing cells of the peripheral blood, and comprise 0.1–0.5% of the circulating population of "nonerythrocytic cells". They were first identified as cells that, upon isolation from blood and subsequent *in vitro* culture, exhibited mixed morphological and molecular characteristics of HSCs, monocytes, and fibroblasts. They are present in wounds, at sites of pathological fibroses, and in the reactive stroma of tumors (Metz 2003; Phillips et al. 2004; Bellini and Mattoli 2007; Wynn 2008). It is not clear whether fibrocytes exist in circulation as such, though it is more likely that they represent the obligate intermediate stage of differentiation into mature mesenchymal cells of a bone marrow-derived precursor of the monocyte lineage that circulates and becomes a "fibrocyte" only at specific tissue sites under permissive conditions (Bellini and Mattoli 2007). Fibrocytes are thought to derive from the hematopoietic lineage since they express cell surface antigens such as CD34, CD45, and CD11b, though they produce matrix proteins such as collagen type I and fibronectin; moreover, their bone marrow origin has been extensively documented in transplant models (Bellini and Mattoli 2007). They also constitutively secrete ECM degrading enzymes, primarily MMP9 (Quan et al. 2004; van Deventer et al. 2008), which promotes endothelial cell invasion, and several proangiogenic factors including VEGF, bFGF, IL-8, and PDGF (Hartlapp et al. 2001; Metz 2003; Quan et al. 2004). Collectively, these findings suggest that the fibroblast population of bone marrow origin contributing to wound healing proposed 130 years ago by the German pathologist Cohneim could indeed be formed, at least in part, by fibrocytes. Of note, although the bone marrow appears to give origin to fibrocyte precursors, to this date the presence of differentiated fibrocytes in the bone marrow has not yet been reported.

2.3   Contribution of the Adult Bone Marrow to the Tumor Stroma

As aforementioned, the contribution of the bone marrow to the reactive stroma of tumors is an active field of investigation, as a tumor is not only composed of transformed cells, but is also intimately associated with endothelial cells, fibroblasts, and inflammatory cells that constitute its stroma and can potentially influence its growth (Hughes 2008; Zumsteg and Christofori 2009). Each of these cell types is contributed, at least in part, by the bone marrow. Bone marrow is the source of TAMs or tumor-associated macrophages (Wels et al. 2008). Accumulation of TAMs in the hypoxic regions of tumors, i.e., in the most malignant ones, has been well documented, and is likely regulated by a hypoxic-mediated chemoattractive gradient involving growth factors such as VEGF, which is induced by the HIF-1 (hypoxia-induced factor-1) transcription factor (Wels et al. 2008). TAMs and inflammation in general are believed to promote tumor development and progression by a variety of mechanisms including angiogenesis and remodeling of the ECM (Coussens et al. 2000; van Kempen and Coussens 2002; Balkwill and Coussens 2004; Mueller and Fusenig 2004; Tan and Coussens 2007; Pahler et al. 2008; DeNardo et al. 2009). VEGF produced by the hypoxic tumor also mobilizes EPCs from the bone marrow, which ultimately infiltrate and get incorporated into the newly formed vasculature (Wels et al. 2008). Lastly, the adult bone marrow provides CAF precursors (Wels et al. 2008), although it is still uncertain whether tumor hypoxia is critical in their recruitment to the tumor.

3 Carcinoma-Associated Fibroblasts

3.1 Definition of CAFs

CAFs or carcinoma-associated fibroblasts are the most abundant cells of the tumor microenvironment (De Wever and Mareel 2003; Orimo et al. 2005; Baglole et al. 2006; Kalluri and Zeisberg 2006; Orimo and Weinberg 2006). CAFs are associated with cancer cells at all stages of cancer progression, and their structural and functional contributions to this process are just beginning to emerge. The vast majority of CAFs are indeed myofibroblasts. Myofibroblasts are typically present at the sites of wound healing and chronic inflammation. As myofibroblasts, CAFs are spindle-shaped mesenchymal cells that share characteristics with both smooth muscle cells and fibroblasts. Their presence is revealed by their expression of alpha-smooth muscle actin (a-SMA), which is a classical marker of smooth muscle cells, and of vimentin, which is a classical marker of fibroblasts (De Wever et al. 2008). Of note, α-SMA is absent in normal dermal fibroblasts (De Wever et al. 2008).

3.2 Tissue-Origin of CAFs and Molecular Mechanisms that Control Their Differentiation and Activity

Since CAFs lack genetic mutations typically found in neighboring tumor cells, a possible tumor origin of these cells following epithelial-to-mesenchymal transition seems to be unlikely (Kalluri and Zeisberg 2006). Conversely, *in vivo* evidence in favor of fibroblast invasion into the tumor compartment has been recently provided (Fukumura et al. 1998). Solid tumor implantation in transgenic mice expressing GFP under the control of the VEGF promoter leads to induction of host VEGF promoter activity in fibroblasts (Fukumura et al. 1998). With time, GFP-positive fibroblast-like cells invade the tumor and can be seen throughout the tumor mass (Fukumura et al. 1998).

CAFs might differentiate from normal stromal fibroblasts of the surrounding tissue, from bone marrow-derived MSCs recruited to the tumor (Hall et al. 2007), or from bone marrow-derived fibrocyte precursors (Ishii et al. 2003; Metz 2003; Direkze et al. 2004, 2006; Bellini and Mattoli 2007) upon activity of growth factors such as TGFβ1. TGFβ1 is a growth factor secreted by a range of tumor cells and is known to mediate the interaction of cancer cells with stromal fibroblasts. In particular, the ability of TGFβ1 to induce fibrocyte differentiation into myofi-broblasts has been extensively documented *in vitro* (Metz 2003; Hong et al. 2007).

Upon TGFβ1 treatment human CD34(+) alphaSMA(−) fibrocytes can differentiate into CD34(−) alpha SMA(+) myofibroblasts in *in vitro* cultures (Hong et al. 2007). *In vivo*, several histopathological studies have correlated the loss of CD34(+) stromal cells at tumor sites with malignant potential, and other studies have associated the loss of CD34 expression with an increase in alpha SMA production (Barth et al. 2004; Quan et al. 2004; Barth and Westhoff 2007). It is possible that each of these changes could be the consequence of fibrocytes differentiating into CAFs. Both cancer cells and cells at the interface of malignant lesions produce high amounts of TGFβ1 and hypoxic regions of tumors contain elevated levels of endothelin-1 (ET-1) (Bellini and Mattoli 2007). It has been suggested the apparent loss of CD34(+) cells and the concomitant increase in the number of CD34(−) alphaSMA(+) cells in the stroma surrounding malignant lesions may indicate an increased differentiation of CD34(+) fibrocytes into mature CD34(−) CAFs triggered by the presence of excessive levels of both TGFβ1 and ET1 (Bellini and Mattoli 2007). Since fibrocytes, but not CAFs, have antigen-presenting capabilities (Bellini and Mattoli 2007), it is tempting to speculate that fibrocytes could be important for local immunosurveillance, whereas their differentiation into CAFs would favor a more invasive phenotype of malignant tumors (Feldon et al. 2006).

3.3 Regulation of Tumor Initiation, Progression and Metastasis by CAFs

Numerous lines of evidence suggest an important role for CAFs in tumor initiation, progression, and metastasis. When normal human prostate epithelial cells immortalized by SV40 T-antigen are introduced into nude mice, they fail to form tumors. However, if the same cells are mixed with CAFs extracted from human prostate carcinoma, tumors arise. Interestingly, tumors do not form if immortalized prostate epithelial cells are mixed with fibroblasts isolated from normal prostate (Olumi et al. 1999). Furthermore, bone-marrow-derived human MSCs, which are thought to contribute to CAFs, when mixed with weakly metastatic human breast carcinoma cells, increase their ability to metastasize upon subcutaneous transplantation, with a mechanism that involves the chemokine CCL5 secreted by the MSCs, and its receptor CCR5 present on cancer cells (Karnoub et al. 2007).

In addition, CAFs extracted from human breast carcinomas promote the growth of mixed breast carcinoma cells significantly more than normal mammary fibroblasts derived from the same patients (Orimo et al. 2005; Orimo and Weinberg 2006). In these experimental conditions, the tumor-promoting effect of CAFs appears to be linked to their ability to secrete stromal cell-derived factor 1 (SDF-1) or CXCL12, which both promotes the recruitment of EPCs from the bone marrow and directly stimulates growth of tumor cells through its action on the cognate receptor CXCR4 expressed on carcinoma cells (Orimo et al. 2005; Orimo and Weinberg 2006).

It has been reported that inactivation of PTEN in stromal fibroblasts of mouse mammary glands favors malignant transformation of mammary epithelial tumors, by mechanisms involving the recruitment of Ets2 to target promoters (Trimboli et al. 2009). Recent experimental evidence has also shown that senescent epithelial cells promote the formation of epithelial tumors by yet unknown mechanisms (Shan et al. 2009). Another example of how the stroma influences tumor progression is provided by studies on the effects of radiotherapy on cancer cells. There is evidence suggesting that the antitumor effect of the radiotherapy is contributed at least in part by the generation of fibrotic scar tissue that restrains tumor invasion. However, clinical and experimental observations also indicate that irradiated stroma might exert tumor-promoting effects (Milas et al. 1988). Tumors growing within a pre-irradiated stroma have reduced growth but show a more invasive and metastatic phenotype, most likely because they have reduced angiogenesis and they are more hypoxic compared to control tumors (Monnier et al. 2008).

The current working hypothesis is that CAFs aid tumorigenesis at least in part through their ability to support vasculogenesis and angiogenesis, with both VEGF dependent and independent mechanisms (Dong et al. 2004). In addition to proangiogenic factors, CAFs also secrete a variety of serine proteases and metalloproteinases, which favor tumor growth, progression, and metastasis by degrading and remodeling the ECM (Masson et al. 1998). In this regard, imaging of co-cultures of squamous cell carcinomas and CAFs revealed that stromal cells direct the migration of epithelial cancer cells by triggering both proteolytic and structural changes of the ECM (Gaggioli et al. 2007). Lastly, CAFs positively modulate the recruitment of inflammatory cells and/or directly stimulate tumor growth and progression by expressing a wide range of growth factors, cytokines, and chemokines (Nazareth et al. 2007; Koyama et al. 2008).

Interestingly, conditional inactivation of the TGFβ type II receptor gene in mouse fibroblasts (Tgfbr2fspKO) led to prostate and stomach cancer, both associated with an increased abundance of stromal cells (Bhowmick et al. 2004; Stover et al. 2007). Activation of paracrine hepatocyte growth factor (HGF) signaling was identified as one possible mechanism for cancer initiation and progression. These findings show a tumor-suppressive role for TGFβ signaling in fibroblasts, in part by suppressing HGF signaling between mammary fibroblasts and epithelial cells, and they highlight the extreme complexity and "bipolar" nature of tumor-stroma cross-talks (Bhowmick et al. 2004; Mueller and Fusenig 2004; Stover et al. 2007).

The notion that not only the stroma influences cancer cells, but also cancer cells affect the stroma, is supported by the recent finding that in a mouse model of prostate carcinoma, cancer cells stimulate the induction of p53 protein in CAFs through a paracrine mechanism. This process generates a selective pressure that leads to the expansion of a subpopulation of CAFs that lack p53 and promote cancer invasion (Hill et al. 2005). Notably, in human breast cancers, loss of p53 in the stroma has been significantly associated with lymph nodes metastasis (Patocs et al. 2007).

4 Tumor Hypoxia and CAFs

4.1 Chronic and Acute Hypoxia in Malignancies

Hypoxia is one of the hallmarks of malignant tumors (Hall and Giaccia 2006). Hypoxia in tumors can result from two different mechanisms: chronic vs. acute. Chronic hypoxia is the consequence of oxygen consumption by tumor cells that are inadequately perfused due to the distance between the cancer cells and the nearest blood vessels. Acute hypoxia is the result of the temporary closing of a blood vessel due to the malformed vasculature of the tumor. Tumor cells are exposed to a continuum of oxygen concentrations, ranging from the highest oxygen tensions surrounding cells nearby capillaries to the lowest oxygen tensions surrounding cells that are the most distant from the capillaries. Both chronic and acute hypoxia are correlated with poor outcome, and have been shown to drive malignant progression. Hypoxia increases the malignancy of cancer cells by gene amplification, genomic instability, and selecting for cells that lack wild-type p53 (Graeber et al. 1996). Moreover, hypoxia promotes angiogenesis through the induction of proangiogenic mitogens such as VEGF, increases glucose uptake and glycolytic metabolism, increases the expression of ABC transporters, promotes invasive growth, and reduces the effectiveness of therapies that require oxygen to be effective such as radiotherapy (Leo et al. 2004; Erler et al. 2006; Koukourakis et al. 2006; Chan et al. 2007; Cosse and Michiels 2008; Rankin and Giaccia 2008; Erler et al. 2009; Semenza 2009).

Up to one percent of the genome is transcriptionally regulated by hypoxic stress (Denko et al. 2003; Giaccia et al. 2004; Chan et al. 2007). A substantial portion of hypoxia-induced genes are regulated by HIF-1. Hypoxia-inducible factor-1 (HIF-1), a ubiquitously expressed transcription factor, is a major regulator of cellular adaptation to hypoxia (Bunn and Poyton 1996; Kaelin 2002; Giaccia et al. 2003; Semenza 2003; Liu and Simon 2004). HIF is a heterodimeric DNA-binding complex that consists of two basic helix-loop-helix (bHLH) proteins of the PER/ARNT/SIM (PAS) subfamily, HIF-1α and HIF-1β (Wang et al. 1995). Interestingly, while HIF-1α and HIF-1β mRNAs are ubiquitously expressed (Wenger et al. 1997), HIF-1α protein is rapidly degraded under normoxic conditions but increases exponentially as O_2 levels fall below 5% (Wang and Semenza 1993; Ivan et al. 2001; Jaakkola et al. 2001; Chan et al. 2002; Min et al. 2002; Pouyssegur et al. 2006). In contrast, HIF-1β (also known as aryl hydrocarbon nuclear translocator or ARNT) is non-oxygen responsive. The HIF-1α/HIF-1β complex binds to a specific sequence 5$'$-RCGTG-3$'$ (where R denotes a purine residue) termed hypoxia response elements (HREs), and transactivates target genes containing HREs (Kallio et al. 1998). HIF-1α itself does not directly sense variations in O_2 tension (Chan et al. 2005). Instead, a class of 2-oxoglutarate-dependent and Fe^{2+}-dependent dioxygenases act as O_2 sensors (Pouyssegur et al. 2006). Prolyl-hydroxylase domain proteins (PHDs) are the O_2 sensors involved in HIF-1α or its family member HIF-2α degradation. PHDs hydroxylate two prolyl residues (P402 and P564) in the HIF-α region referred to as the O_2-dependent degradation domain (ODD) (Berra et al. 2003). This modification occurs in normoxic conditions and mediates the binding of the von Hippel-Lindau tumor suppressor protein (pVHL) – which is an E3 ubiquitin ligase – to the alpha subunit of HIF-1 or HIF-2. These subunits are then marked with polyubiquitin chains and targeted for degradation by the proteasome. Under hypoxic conditions, the activity of the PHDs decreases, resulting in diminished proline hydroxylation. As a result, HIF-1α or HIF-1β protein accumulates, translocates to the nucleus, dimerizes with HIF-1β, recruits transcriptional co-activators, and binds to HREs within the promoters of hypoxia-responsive genes (Kallio et al. 1999). To date, more than one hundred putative HIF target genes have been identified (Wykoff et al. 2000; Bishop et al. 2004; Leo et al. 2004; Greijer et al. 2005). They are involved in a wide variety of biological processes including energy metabolism, angiogenesis, erythropoiesis, cell survival, apoptosis, redox, and pH regulation (Maxwell et al. 1999; Greijer et al. 2005).

Several other transcription factors are also activated by hypoxia: cyclic-AMP response element binding (CREB) protein, the activator protein-1 (AP1), the nuclear factor kb (NF-kb), and the early growth response-1 protein (Egr-1). Moreover, other important non-HIF mediated hypoxia-signaling pathways involve the unfolded protein response (UPR) and the mTOR (target of rapamycin) kinase signaling pathway (Wouters and Koritzinsky 2008) (Feldman et al. 2005).

4.2   CAFs and Hypoxia: A Working Hypothesis

As discussed above, the role of hypoxia in recruitment of TAMs and EPCs has been previously documented. Although CAFs and hypoxia are both crucial for tumor progression, little is known about how hypoxia affects recruitment and activation of CAFs. Experimental evidence suggests that hypoxia might regulate differentiation and activity of CAFs in malignant tumors in a variety of different ways (Fig. 1).

Hypoxia and the HIF family of transcription factors have been linked to fibrosis in a variety of tissues and during various pathological conditions. HIF-1 promotes proximal renal fibrosis by regulating the endothelial-mesenchymal transition in the kidney (Higgins et al. 2008). Moreover, hypoxia and HIFs drive fibrosis by controlling the remodeling of the ECM and modulating the TGFβ signaling pathway (Higgins et al. 2008). Lastly, experimental evidence indicates that recruitment of fibrocytes/myofibroblasts to sites of pathological fibroses may be driven by hypoxia (Phillips et al. 2004; Hayashida et al. 2005; Jiang et al. 2006; Mehrad et al. 2009). Although it is unknown whether tumor hypoxia does indeed play a role in recruitment of CAF precursors to the tumor, tumor hypoxia may control differentiation of CAFs from precursor cells through mechanisms involving TGFβ and ET-1, which have been shown to be important for myofibroblast formation and are known to be regulated by hypoxia (see above).

Moreover, it is now established that CAFs promote angiogenesis by releasing chemotactic signals such as CXCL12, which help to recruit EPCs into the tumor

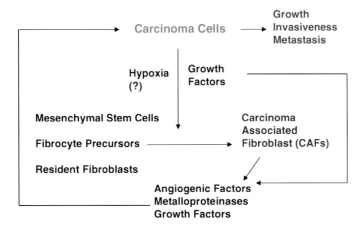

Fig. 1 Carcinoma-associated fibroblasts (CAFs) play a major role in organizing the tumor microenvironment and in modulating tumor growth, invasion, and metastasis. For more details, see text

stroma, and by secreting a series of proangiogenic factor such VEGF, BFGF, IL8, and angiopoietins. Notably, each of these agents is a well-known downstream target of hypoxia, and increased angiogenesis is one of the hallmarks of the hypoxic response. Lastly, the selective pressure for the expansion of a subpopulation of CAFs that lack p53, as it has been documented both in experimental models of cancer and in human cancers (Hill et al. 2005; Patocs et al. 2007), could be indeed the consequence of the hypoxic microenvironment of malignant tumors.

Paradoxically, CAFs, at least in some settings, may aid the survival of hypoxic cancer cells by undergoing metabolic changes that mirror the phenotype of malignant cells. Immunohistochemical analysis of colorectal adenocarcinomas has demonstrated that cancer cells express enzyme/transporter activities suggestive of anaerobic metabolism. In contrast, CAFs express proteins involved in lactate absorption and oxidation, concomitant with reduced glucose absorption. Overall, the findings are consistent with the notion that CAFs express complementary metabolic pathways, buffering and recycling products of anaerobic metabolism to promote cancer cell survival (Koukourakis et al. 2006).

5 Future Perspectives

As we have emphasized in this brief summary, a variety of experimental evidence supports the notion that CAFs are likely to be crucial for tumor progression and metastasis, and that targeting CAFs and the microenvironment could be a promising approach to cancer therapy (Ahmed et al. 2008; Kiaris et al. 2008; Aharinejad et al. 2009; Anton and Glod 2009; Chometon and Jendrossek 2009). However, much needs still to be discovered about these cells and their functional and structural contributions to cancer. A series of critical questions related to the origin of CAFs and to modulation of their activity need to be addressed. How does regulation of bone marrow homeostasis *in vivo* affect the number and activity of CAFs in carcinomas? Is it possible to control the CAF population by altering the remodeling process in bone and in the bone marrow? How much of the genetic make-up of CAFs is the consequence of their adaptation to hypoxia? Is there a "CAF signature" across fibroblasts isolated from a variety of tumors with different degrees of malignancy? The ultimate goal of addressing these questions would be the identification of pharmacological tools that could be used to control the CAF population *in vivo* and its effects on tumor progression (Ahmed et al. 2008; Kiaris et al. 2008; Aharinejad et al. 2009; Anton and Glod 2009; Chometon and Jendrossek 2009).

To this end, precursors of CAFs could be used as a vehicle for anticancer therapy. It has been suggested that MSCs home to tumors and contribute to CAFs. Theoretically, MSCs could thus be genetically modified to produce anticancer agents to be delivered in situ. This is an exciting possibility that should be considered as a new approach to cancer therapy (Hall et al. 2007).

References

Aharinejad S, Sioud M et al (2009) Targeting stromal-cancer cell interactions with siRNAs. Methods Mol Biol 487:243–266

Ahmed F, Steele JC et al (2008) Tumor stroma as a target in cancer. Curr Cancer Drug Targets 86:447–453

Albini A, Sporn MB (2007) The tumour microenvironment as a target for chemoprevention. Nat Rev Cancer 72:139–147

Anton K, Glod J (2009) Targeting the tumor stroma in cancer therapy. Curr Pharm Biotechnol 102:185–191

Asahara T, Takahashi T et al (1999) VEGF contributes to postnatal vascularization by mobilizing bone marrow-derived endothelial progenitor cells. EMBO J 18:3964–3972

Baglole CJ, Ray DM et al (2006) More than structural cells, fibroblasts create and orchestrate the tumor microenvironment. Immunol Invest 353–4:297–325

Balkwill F, Coussens LM (2004) Cancer: an inflammatory link. Nature 4317007:405–406

Balkwill F, Mantovani A (2001) Inflammation and cancer: back to Virchow? Lancet 3579255:539–545

Barth PJ, Westhoff CC (2007) CD34+ fibrocytes: morphology, histogenesis and function. Curr Stem Cell Res Ther 23:221–227

Barth PJ, Schenck zu Schweinsberg T et al (2004) CD34+ fibrocytes, alpha-smooth muscle antigen-positive myofibroblasts, and CD117 expression in the stroma of invasive squamous cell carcinomas of the oral cavity, pharynx, and larynx. Virchows Arch 4443:231–234

Bellini A, Mattoli S (2007) The role of the fibrocyte, a bone marrow-derived mesenchymal progenitor, in reactive and reparative fibroses. Lab Invest 879:858–870

Berra E, Benizri E et al (2003) HIF prolyl-hydroxylase 2 is the key oxygen sensor setting low steady-state levels of HIF-1alpha in normoxia. Embo J 2216:4082–4090

Bhowmick NA, Chytil A et al (2004) TGF-beta signaling in fibroblasts modulates the oncogenic potential of adjacent epithelia. Science 3035659:848–851

Bianco P, Robey P et al (2008) Mesenchymal stem cells: revisiting history, concepts and assays. Cell Stem Cell 2:313–319

Bishop T, Lau K et al (2004) Genetic analysis of pathways regulated by the von hippel-lindau tumor suppressor in Caenorhabditis elegans. PLoS Biol 2:e289

Bissell MJ, Radisky DC et al (2002) The organizing principle: microenvironmental influences in the normal and malignant breast. Differentiation 709–10:537–546

Bucala R, Spiegel LA et al (1994) Circulating fibrocytes define a new leukocyte subpopulation that mediates tissue repair. Mol Med 11:71–81

Bunn HF, Poyton RO (1996) Oxygen sensing and molecular adaptation to hypoxia. Physiol Rev 763:839–885

Caplan A (2007) Adult mesenchymal stem cells for tissue engineering versus regenerative medicine. J Cell Physiol 213:341–347

Chan D, Suthphin P et al (2002) Role of prolyl hydroxylation in oncogenically stabilized hypoxia-inducible factor-1alpha. J Biol Chem 277:40112–40117

Chan D, Sutphin P et al (2005) Coordinate regulation of the oxygen-dependent degradation domains of hypoxia-inducible factor 1 alpha. Mol Cell Biol 25:6415–6426

Chan DA, Krieg AJ et al (2007) HIF gene expression in cancer therapy. Methods Enzymol 435:323–345

Chometon G, Jendrossek V (2009) Targeting the tumour stroma to increase efficacy of chemo- and radiotherapy. Clin Transl Oncol 112:75–81

Cosse JP, Michiels C (2008) Tumour hypoxia affects the responsiveness of cancer cells to chemotherapy and promotes cancer progression. Anticancer Agents Med Chem 87:790–797

Coussens LM, Werb Z (2002) Inflammation and cancer. Nature 4206917:860–867

Coussens LM, Tinkle CL et al (2000) MMP-9 supplied by bone marrow-derived cells contributes to skin carcinogenesis. Cell 1033:481–490

De Wever O, Mareel M (2003) Role of tissue stroma in cancer cell invasion. J Pathol 2004:429–447

De Wever O, Demetter P et al (2008) Stromal myofibroblasts are drivers of invasive cancer growth. Int J Cancer 12310:2229–2238

DeNardo DG, Barreto JB et al (2009) CD4(+) T cells regulate pulmonary metastasis of mammary carcinomas by enhancing protumor properties of macrophages. Cancer Cell 162:91–102

Denko N, Fontana L et al (2003) Investigating hypoxic tumor physiology through gene expression patterns. Oncogene 22:5907–5914

Direkze NC, Hodivala-Dilke K et al (2004) Bone marrow contribution to tumor-associated myofibroblasts and fibroblasts. Cancer Res 6423:8492–8495

Direkze NC, Jeffery R et al (2006) Bone marrow-derived stromal cells express lineage-related messenger RNA species. Cancer Res 663:1265–1269

Dong J, Grunstein J et al (2004) VEGF-null cells require PDGFR alpha signaling-mediated stromal fibroblast recruitment for tumorigenesis. Embo J 2314:2800–2810

Dvorak HF (1986) Tumors: wounds that do not heal. Similarities between tumor stroma generation and wound healing. N Engl J Med 31526:1650–1659

Erler J, Bennewith K et al (2006) Lysyl oxidase is essential for hypoxia-induced metastasis. Nature 440:1222–1226

Erler JT, Bennewith KL et al (2009) Hypoxia-induced lysyl oxidase is a critical mediator of bone marrow cell recruitment to form the premetastatic niche. Cancer Cell 151:35–44

Feldman DE, Chauhan V et al (2005) The unfolded protein response: a novel component of the hypoxic stress response in tumors. Mol Cancer Res 311:597–605

Feldon SE, O'Loughlin CW et al (2006) Activated human T lymphocytes express cyclooxygenase-2 and produce proadipogenic prostaglandins that drive human orbital fibroblast differentiation to adipocytes. Am J Pathol 1694:1183–1193

Friedenstein A (1980) Stromal mechanisms of bone marrow: cloning *in vitro* and retransplantation *in vivo*. Haematol Blood Transf 25:19–29

Friedenstein A, Chailakhjan R et al (1970) The development of fibroblast colonies in monolayer cutures of guinea-pig bone marrow and spleen cells. Cell Tissue Kinet 3:393–403

Fukumura D, Xavier R et al (1998) Tumor induction of VEGF promoter activity in stromal cells. Cell 946:715–725

Gaggioli C, Hooper S et al (2007) Fibroblast-led collective invasion of carcinoma cells with differing roles for RhoGTPases in leading and following cells. Nat Cell Biol 912:1392–1400

Giaccia A, Siim B et al (2003) HIF-1 as a target for drug development. Nat Rev Drug Discov 2:803–811

Giaccia AJ, Simon MC et al (2004) The biology of hypoxia: the role of oxygen sensing in development, normal function, and disease. Genes Dev 1818:2183–2194

Graeber TG, Osmanian C et al (1996) Hypoxia-mediated selection of cells with diminished apoptotic potential in solid tumours. Nature 3796560:88–91

Greijer AE, van der Groep P et al (2005) Up-regulation of gene expression by hypoxia is mediated predominantly by hypoxia-inducible factor 1 (HIF-1). J Pathol 2063:291–304

Hall E, Giaccia AJ (2006) Radiobiology for the radiologist. Lippincott Williams & Wilkins, Philadelphia

Hall B, Andreeff M et al (2007) The participation of mesenchymal stem cells in tumor stroma formation and their application as targeted-gene delivery vehicles. Handb Exp Pharmacol 180:263–283

Hanahan D, Weinberg RA (2000) The hallmarks of cancer. Cell 1001:57–70

Hartlapp I, Abe R et al (2001) Fibrocytes induce an angiogenic phenotype in cultured endothelial cells and promote angiogenesis *in vivo*. Faseb J 1512:2215–2224

Hayashida K, Fujita J et al (2005) Bone marrow-derived cells contribute to pulmonary vascular remodeling in hypoxia-induced pulmonary hypertension. Chest 1275:1793–1798

Higgins DF, Kimura K et al (2008) Hypoxia-inducible factor signaling in the development of tissue fibrosis. Cell Cycle 79:1128–1132

Hill R, Song Y et al (2005) Selective evolution of stromal mesenchyme with p53 loss in response to epithelial tumorigenesis. Cell 1236:1001–1011

Hong KM, Belperio JA et al (2007) Differentiation of human circulating fibrocytes as mediated by transforming growth factor-beta and peroxisome proliferator-activated receptor gamma. J Biol Chem 28231:22910–22920

Hughes CC (2008) Endothelial-stromal interactions in angiogenesis. Curr Opin Hematol 153:204–209

Ishii G, Sangai T et al (2003) Bone-marrow-derived myofibroblasts contribute to the cancer-induced stromal reaction. Biochem Biophys Res Commun 3091:232–240

Ivan M, Kondo K et al (2001) HIFalpha targeted for VHL-mediated destruction by proline hydroxylation: imlications for O2 sensing. Science 292:464–468

Jaakkola P, Mole D et al (2001) Targeting of HIF-alpha to the von Hippel-Lindau ubiquitylation complex by O2-regulated prolyl hydroxylation. Science 292:468–472

Jiang YL, Dai AG et al (2006) Transforming growth factor-beta1 induces transdifferentiation of fibroblasts into myofibroblasts in hypoxic pulmonary vascular remodeling. Acta Biochim Biophys Sin (Shanghai) 381:29–36

Kaelin W (2002) How oxygen makes its presence felt. Genes Dev 16:1441–1445

Kallio PJ, Okamoto K et al (1998) Signal transduction in hypoxic cells: inducible nuclear translocation and recruitment of the CBP/p300 coactivator by the hypoxia-inducible factor-1alpha. Embo J 1722:6573–6586

Kallio PJ, Wilson WJ et al (1999) Regulation of the Hypoxia-inducible transcription factor 1alpha by the ubiquitin-proteasome pathway. J Biol Chem 27410:6519–6525

Kalluri R, Zeisberg M (2006) Fibroblasts in cancer. Nat Rev Cancer 65:392–401

Karnoub AE, Dash AB et al (2007) Mesenchymal stem cells within tumour stroma promote breast cancer metastasis. Nature 4497162:557–563

Kiaris H, Trimis G et al (2008) Regulation of tumor-stromal fibroblast interactions: implications in anticancer therapy. Curr Med Chem 1529:3062–3067

Kolf C, Cho E et al (2007) Biology of adult mesenchymal stem cells: regulation of niche, self-renewal and differentiation. Arthritis Res Ther 9:204

Koukourakis MI, Giatromanolaki A et al (2006) Comparison of metabolic pathways between cancer cells and stromal cells in colorectal carcinomas: a metabolic survival role for tumor-associated stroma. Cancer Res 662:632–637

Koyama H, Kobayashi N et al (2008) Significance of tumor-associated stroma in promotion of intratumoral lymphangiogenesis: pivotal role of a hyaluronan-rich tumor microenvironment. Am J Pathol 1721:179–193

Leo C, Giaccia A et al (2004) The hypoxic tumor microenvironment and gene expression. Semin Radiat Oncol 14:207–214

Liotta LA, Kohn EC (2001) The microenvironment of the tumour-host interface. Nature 4116835:375–379

Liu L, Simon MC (2004) Regulation of transcription and translation by hypoxia. Cancer Biol Ther 36:492–497

Lorusso G, Ruegg C (2008) The tumor microenvironment and its contribution to tumor evolution toward metastasis. Histochem Cell Biol 1306:1091–1103

Masson R, Lefebvre O et al (1998) *In vivo* evidence that the stromelysin-3 metalloproteinase contributes in a paracrine manner to epithelial cell malignancy. J Cell Biol 1406:1535–1541

Maxwell P, Wiesener M et al (1999) The tumour suppressor protein VHL targets hypoxia-inducible factors for oxygen-dependent proteolysis. Nature 399:271–275

Mehrad B, Burdick MD et al (2009) Fibrocyte CXCR4 regulation as a therapeutic target in pulmonary fibrosis. Int J Biochem Cell Biol 418–9:1708–1718

Metz CN (2003) Fibrocytes: a unique cell population implicated in wound healing. Cell Mol Life Sci 607:1342–1350

Milas L, Hirata H et al (1988) Effect of radiation-induced injury of tumor bed stroma on metastatic spread of murine sarcomas and carcinomas. Cancer Res 488:2116–2120

Min J, Yang H et al (2002) Structure of an HIF-1alpha-pVHL complex: hydroxyproline recognition in signaling. Science 296:1886–1889

Monnier Y, Farmer P et al (2008) CYR61 and alphaVbeta5 integrin cooperate to promote invasion and metastasis of tumors growing in preirradiated stroma. Cancer Res 6818:7323–7331

Mueller MM, Fusenig NE (2004) Friends or foes – bipolar effects of the tumour stroma in cancer. Nat Rev Cancer 411:839–849

Nazareth MR, Broderick L et al (2007) Characterization of human lung tumor-associated fibroblasts and their ability to modulate the activation of tumor-associated T cells. J Immunol 1789:5552–5562

Olumi AF, Grossfeld GD et al (1999) Carcinoma-associated fibroblasts direct tumor progression of initiated human prostatic epithelium. Cancer Res 5919:5002–5011

Orimo A, Weinberg RA (2006) Stromal fibroblasts in cancer: a novel tumor-promoting cell type. Cell Cycle 515:1597–1601

Orimo A, Gupta PB et al (2005) Stromal fibroblasts present in invasive human breast carcinomas promote tumor growth and angiogenesis through elevated SDF-1/CXCL12 secretion. Cell 1213:335–348

Pahler JC, Tazzyman S et al (2008) Plasticity in tumor-promoting inflammation: impairment of macrophage recruitment evokes a compensatory neutrophil response. Neoplasia 104:329–340

Patocs A, Zhang L et al (2007) Breast-cancer stromal cells with TP53 mutations and nodal metastases. N Engl J Med 35725:2543–2551

Phillips RJ, Burdick MD et al (2004) Circulating fibrocytes traffic to the lungs in response to CXCL12 and mediate fibrosis. J Clin Invest 1143:438–446

Pietras K, Pahler J et al (2008) Functions of paracrine PDGF signaling in the proangiogenic tumor stroma revealed by pharmacological targeting. PLoS Med 51:e19

Pittenger M, Mackay A et al (1999) Multiineage potential of adult human mesenchymal stem cells. Science 284:143–147

Pouyssegur J, Dayan F et al (2006) Hypoxia signalling in cancer and approaches to enforce tumour regression. Nature 4417092:437–443

Prockop D (1997) Marrow stromal cells as stem cells for nonhematopoietic tissues. Science 276:71–74

Quan TE, Bucala R (2007) Culture and analysis of circulating fibrocytes. Methods Mol Med 135:423–434

Quan TE, Cowper S et al (2004) Circulating fibrocytes: collagen-secreting cells of the peripheral blood. Int J Biochem Cell Biol 364:598–606

Rankin EB, Giaccia AJ (2008) The role of hypoxia-inducible factors in tumorigenesis. Cell Death Differ 154:678–685

Schipani E, Kronenberg HM (2008) Adult mesenchymal stem cells. Stembook doi:10.3824/stembook.1.38.1

Semenza GL (2003) Targeting HIF-1 for cancer therapy. Nat Rev Cancer 310:721–732

Semenza GL (2009) Regulation of cancer cell metabolism by hypoxia-inducible factor 1. Semin Cancer Biol 191:12–16

Shan W, Yang G et al (2009) The inflammatory network: bridging senescent stroma and epithelial tumorigenesis. Front Biosci 14:4044–4057

Stover DG, Bierie B et al (2007) A delicate balance: TGF-beta and the tumor microenvironment. J Cell Biochem 1014:851–861

Tan TT, Coussens LM (2007) Humoral immunity, inflammation and cancer. Curr Opin Immunol 192:209–216

Tlsty TD, Coussens LM (2006) Tumor stroma and regulation of cancer development. Annu Rev Pathol 1:119–150

Trimboli AJ, Cantemir-Stone CZ et al (2009) Pten in stromal fibroblasts suppresses mammary epithelial tumours. Nature 4617267:1084–1091

van Deventer HW, Wu QP et al (2008) C-C chemokine receptor 5 on pulmonary fibrocytes facilitates migration and promotes metastasis via matrix metalloproteinase 9. Am J Pathol 1731:253–264

van Kempen LC, Coussens LM (2002) MMP9 potentiates pulmonary metastasis formation. Cancer Cell 24:251–252

Wang GL, Semenza GL (1993) General involvement of hypoxia-inducible factor 1 in transcriptional response to hypoxia. Proc Natl Acad Sci USA 909:4304–4308

Wang GL, Jiang BH et al (1995) Hypoxia-inducible factor 1 is a basic-helix-loop-helix-PAS heterodimer regulated by cellular O2 tension. Proc Natl Acad Sci USA 9212:5510–5514

Weinberg RA (2007) The Biology of Cancer 1st Edition Chapter 13:527–586. Garland Science Publisher, NY, USA

Wels J, Kaplan RN et al (2008) Migratory neighbors and distant invaders: tumor-associated niche cells. Genes Dev 225:559–574

Wenger RH, Rolfs A et al (1997) The mouse gene for hypoxia-inducible factor-1alpha–genomic organization, expression and characterization of an alternative first exon and 5' flanking sequence. Eur J Biochem 2461:155–165

Wouters BG, Koritzinsky M (2008) Hypoxia signalling through mTOR and the unfolded protein response in cancer. Nat Rev Cancer 811:851–864

Wykoff C, Pugh C et al (2000) Identification of novel hypoxia dependent and independent target genes of the von Hippel Lindau (VHL) tumor suppressor by mRNA differential expression profiling. Oncogene 19:6297–6305

Wynn TA (2008) Cellular and molecular mechanisms of fibrosis. J Pathol 2142:199–210

Zumsteg A, Christofori G (2009) Corrupt policemen: inflammatory cells promote tumor angiogenesis. Curr Opin Oncol 211:60–70

The Role of Hypoxia Regulated microRNAs in Cancer

Robert McCormick, Francesca M. Buffa, Jiannis Ragoussis, and Adrian L. Harris

Contents

Abstract The molecular response of cancer cells to hypoxia is the focus of intense research. In the last decade, research into microRNAs (miRNAs), small RNAs which have a role in regulation of mRNA and translation, has grown exponentially. miR-210 has emerged as the predominant miRNA regulated by hypoxia. Elucidation of its targets points to a variety of roles for this, and other hypoxia-regulated miRNAs (HRMs), in tumour growth and survival. miR-210 expression correlates

R. McCormick, F.M. Buffa, J. Ragoussis, and A.L. Harris (✉)
Molecular Oncology Laboratories, Weatherall Institute of Molecular Medicine, University of Oxford, John Radcliffe Hospital, Oxford OX3 9DS, UK
e-mal: aharris.lab@imm.ox.ac.uk

M. Celeste Simon (ed.), *Diverse Effects of Hypoxia on Tumor Progression*,
Current Topics in Microbiology and Immunology 345, DOI 10.1007/82_2010_76
© Springer-Verlag Berlin Heidelberg 2010, published online: 10 June 2010

with poor survival in cancer patients, and shows promise for future use as a tumour marker or therapeutic agent. The role of miR-210 and other HRMs in cancer biology is the subject of this review.

Abbreviations

Ago2	Argonaute 2
ARE	AU-rich element
CaCo	Colon cancer cell line
CaLu	Lung cancer cell line
CNE	Nasopharyngeal carcinoma epitheliod cell line
DFOM	Desferrioxamine
DRFS	Distant Relapse Free Survival
eIF	Eukaryotic translation initiation factor
Fe–S	Iron sulphur clusters
HRM	Hypoxia-regulated miRNA
HUVEC	Human umbilical vein endothelial cells
IP	Ischaemic preconditioning
IRE	Iron-responsive element
IRES	Internal ribosomal entry sites
miRNA	microRNA
miRNP	Microribonucleoprotein complex
mRNA	messengerRNA
MSC	Mesenchymal stem cell
PABP	Poly-A tail binding protein
P-body	Processing body
qPCR	Quantitative real time polymerase chain reaction
RISC	RNA-induced silencing complex
RNAi/siRNA RNA	Interference/small interfering RNA
UTR	Untranslated region

1 Discovery of miRNAs

miRNAs were first discovered in 1993, in a work on the worm *Caenorhabidititis elegans*. Expression of a 22-nucleotide molecule, lin-4, with antisense complementarity to the $3'$-untranslated region (UTR) of the lin-14 gene, was found to be necessary for progression of worms from the first to second larval stages (Lee et al. 1993; Wightman et al. 1993). Another small RNA molecule, let-7, with critical function in larval development was later discovered. Mutations of this gene were associated with heterochronic developmental defects, due to the dysregulation of the lin-41 gene. Genetic analysis revealed that let-7 was an untranslated RNA, which was complementary to part of the 3′-UTR of the lin-41 gene. It was proposed that lin-4 and let-7 were negatively regulating the lin-4 and lin-41 genes, respectively, by inhibition of the mRNA translation mechanism (Reinhart et al. 2000; Slack et al. 2000).

2 miRNA Structure, Processing and Function

miRNAs are a family of small, non-coding RNA sequences 18–25 nucleotides long. They are transcribed as primary transcripts (pri-miRNA) embedded in either independent non-coding RNAs or in the introns of protein-coding genes. Some miRNAs, such as the miR-17 family, are clustered in polycistronic transcripts to allow coordinated expression (Hayashita et al. 2005). Since the discovery of lin-4 and let-7, the list of known miRNAs has grown to many hundreds, and many more are predicted to exist (Pillai et al. 2007).

Pri-miRNA is processed in the nucleus by Drosha (an RNAse III enzyme) associated with the DGCR8 (DiGeorge syndrome critical region gene 8) double-stranded RNA-binding protein. Cleavage of the pri-miRNA by Drosha results in a molecule about 70 nucleotides in length, termed the pre-miRNA, which translocates to the cytoplasm assisted by the nuclear export factor Exportin 5/RAN GTP. Here, it is further cleaved by a second RNAse III enzyme, Dicer. One strand from the processed double-stranded molecule (the one least stably paired at the 5′ end) enters the RNA-induced silencing complex (RISC) with Argonaute 2 (Ago2) proteins, and the remaining (passenger) strand is degraded. The Ago2 proteins are the effector molecules in the RNA interference (RNAi) mechanism. They may also have a role in processing and cleavage of pre-miRNA in addition to Dicer, whilst facilitating removal of the passenger strand from the RISC complex (Diederichs and Haber 2007) (Fig. 1).

2.1 *Translational Regulation and mRNA Degradation*

The mature miRNA/RISC complex acts as a regulator of gene translation through binding to complementary sites in the 3′-UTR of mRNAs. If the complementarity is perfect, the binding of the miRNA to the mRNA results in cleavage and degradation of the mRNA molecule. This mechanism is thought to be more prevalent in plants, and in animals the degree of complementarity is usually imperfect.

Efficient binding and silencing activity of a miRNA requires exact binding of positions 2–7 of the miRNA, known as the seed region, to the complementary sequence on the target mRNA molecule. Perfectly matched complementary binding of the seed region alone, however, does not accurately predict target repression. Positions 12–17, towards the 3′ end of the miRNA, are often highly conserved,

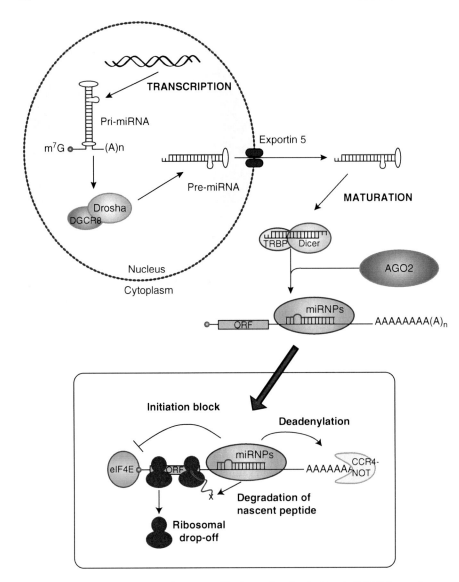

Fig. 1 The pri-miRNA transcript is processed in the nucleus by the Drosha/DGCR8 complex to a ~70-nucleotide pre-miRNA. It is exported in to the cytoplasm by Exportin 5, where it is further cleaved by Dicer/TRBP. One strand of the double-stranded RNA molecule becomes associated with the Ago2/miRNP complex, acting as a guide strand. Binding of the miRNP to a complementary region on the target mRNA leads to inhibition of translation or degradation of the mRNA molecule

and may be important in determining specific target interaction. Additionally, the sequences surrounding the miRNA-responsive element could affect miRNA binding or silencing efficacy (Doench and Sharp 2004; Brennecke et al. 2005;

Grimson et al. 2007). Despite years of refinements to computational target prediction algorithms, it is still an inexact science, and with significant variability in predicted targets between different algorithms. This fact is perhaps reflected in our limited understanding of the mechanism in which miRNAs and their associated protein complex (together known as the microribonucleoprotein complex, or miRNP) lead to gene silencing. Pioneering studies in *C. elegans* demonstrated that lin-4 inhibited lin-14 protein expression, while mRNA levels remained relatively constant, suggesting a post-transcriptional mechanism of repression. They found that the lin-14 3′-UTR was necessary and sufficient to enable repression of the lin-14 protein (Wightman et al. 1993). Since the first studies of miRNAs, increasing evidence has emerged pointing to the role of these molecules in both translational repression and mRNA degradation.

2.2 *Translational Inhibition*

Theories on translational inhibition can broadly be categorised into those describing blocking of translation *initiation,* and those describing repression of translation post-initiation. Before describing these theories, a brief description of the steps of mRNA translation, namely initiation, elongation, and termination, is merited. Initiation occurs with binding of eIF4E, a subunit of the eukaryotic translation initiation factor (eIF) eIF4F, to the 5′-mRNA cap structure (m^7GpppN). eIF4G, another component of the eIF4F complex, binds to the poly-A tail binding protein (PABP1), forming a circular mRNA structure. eIF4G simultaneously binds eIF3 to facilitate assembly of the ribosome complex. The participation of at least ten initiation factors is required to instigate protein translation. Following the assembly of the 40S and 60S ribosomal subunits at an initiation codon (AUG), translation and polypeptide elongation continues until a termination codon is encountered. This results in disassembly of the translational machinery. The circularisation of the mRNA is thought to aid in reassembly of the translational machinery at the 5′ end of the molecule, in readiness to begin the synthesis of another polypeptide molecule (Kapp and Lorsch 2004).

Investigations demonstrating that miRNA-mediated repression occurs exclusively in mRNAs with m^7G cap structure, but not internal ribosomal entry sites (IRES), suggested that repression is occurring at the initiation step (Pillai et al. 2005). In support of this were studies using polysome gradients. It was shown that miRNA-repressed mRNAs sedimented closer to the top of the gradient, indicating reduced ribosomal binding (Pillai et al. 2005). However, another group subsequently demonstrated cap-independent miRNA repression, and suggested that ribosomal drop-off was a more likely mechanistic explanation (Petersen et al. 2006). Notrott et al. showed repression of protein production on actively translating polysomes by let-7, and concluded that the miRNA was interfering with the accumulation of growing peptides (Nottrott et al. 2006). Wakiyama et al. showed miRNA-mediated deadenylation of mRNA, and suggested that shortening of the poly-A tail would prevent PABP1 association and binding to the cap structure, thus preventing translation initiation (Wakiyama et al. 2007). Kiriakidou et al. reported that the Ago proteins contain a limited cap-recognition motif, and argue that competition with eIF4E for binding of the cap would inhibit translational inhibition (Kiriakidou et al. 2007). Against this argument is the fact that Ago would appear to have a lower binding affinity for the cap than eIF4E, and more recent evidence showing that the Ago structure contains amino acids that are incompatible with eIF4E-like mRNA-binding (Kinch and Grishin 2009).

A number of studies have reported co-sedimentation of miRNA with polysomes, and have argued that this supports a post-initiation inhibitory mechanism. However, it has been pointed out that binding of single miRNP to an mRNA will frequently have no significant repressive effect. Co-sedimentation of the miRNPs with polysomes may therefore represent bound miRNAs exerting little or no repressive effect, and therefore cannot be taken as proof of a post-initiation mechanism (Filipowicz et al. 2008).

2.3   mRNA Degradation and P-Bodies

Microarray analyses of cells transfected with synthetic miRNA have demonstrated that mRNA expression of many genes is downregulated as a result, suggesting that degradation or sequestration of the transcripts must be occurring. In addition, analysis of the significantly downregulated transcripts indicates that their 3'-UTRs are enriched with complementary seed-binding regions to the miRNA under investigation (Bartel and Chen 2004; Bagga et al. 2005; Lim et al. 2005).

Processing bodies contain a number of proteins that are involved in mRNA degradation. These include the 5' decapping enzymes DCP1 and DCP2, exoribonuclease 1 (XRN1) which causes 5'–3' decay, and deadenylating enzymes (CCR4-CAF1-NOT complex). Ribosomes and other factors which are required for translation initiation are notably absent, with the exception of eIF4E and eIF4E-transporter, which is implicated in translational repression (Eulalio et al. 2007). Argonaute, miRNAs and their mRNA targets have all been shown to localise to P-bodies. miRNA-mediated mRNA decay has also been shown to be dependent on the components mentioned above, in addition to the GW182 protein (so-called because it contains glycine and typtophan repeats, and because of its molecular weight) (Jakymiw et al. 2005; Liu et al. 2005; Pillai et al. 2005; Meister 2007). While this provides compelling evidence that miRNA-mediated mRNA decay requires the presence of P-body components, it is not clear whether the physical environment of P-bodies is required. This is due to the fact that depletion of certain proteins, e.g., LSm1 (part of the decapping co-activator complex) results in diffusion of Ago2 throughout the cytoplasm, but with no impairment of miRNA-mediated mRNA degradation activity. However, depleting cells of RCK/p54 (a DEAD box helicase) results in P-body loss and relief of miRNA-mediated translational repression (Chu and Rana 2006). Treatment of cells with cyclohexamide stabilises mRNAs into polysomes, and leads to loss of P-bodies. Conversely, inhibition of translation initiation by puromycin, for example, leads to accumulation of P-bodies. This supports the hypothesis that P-bodies are the result of non-translating mRNP accumulation (Anderson and Kedersha 2006). In other words, P-bodies may well be the consequence of miRNA-mediated mRNA repression, rather than the cause of it.

2.4 Translational Stimulation by miRNAs

AU-rich elements (AREs) appear in the $3'$-UTR of 12% of mammalian genes. ARE-binding proteins are known to target these transcripts for exonucleolytic decay of the poly-A tail, leading to mRNA degradation. It was discovered, however, that in the state of cell-cycle quiescence, the presence of AREs can lead to translational stimulation. Studies using reporter assays containing the highly conserved $3'$-UTR ARE of Tumour Necrosis Factor α (TNFα) revealed that in conditions of serum deprivation (leading to quiescence), translation was increased fivefold. Fragile X mental retardation protein 1 (FXR1) was found to be necessary for translational up-regulation, in addition to Ago2. Further studies revealed that a miRNA, miR-369-3p, was essential for TNFα ARE-directed translational up-regulation. Additionally, it was found that miR-369-3p switched from activation of translation (in G0/G1 arrest – quiescence) to repression during the cell cycle, with optimal repression occurring in late S-phase, close to the G2 boundary (Vasudevan and Steitz 2007; Vasudevan et al. 2007, 2008). Cell cycle dependence of miRNA activity could prove to be highly relevant in the study of tumour biology. Fluctuating vascular supply leading to hypoxia and serum deprivation will likely lead to areas of cellular quiescence, and thus variable responses to miRNAs.

3 miRNAs are Dysregulated in Cancer

Early evidence of the significance of miRNA in cancer was demonstrated in the Bantam miRNA, which caused overgrowth of wing and eye tissue in Drosophila through negative regulation of the pro-apoptotic gene hid. Another miRNA, miR-14, was demonstrated to suppress cell death (Hwang and Mendell 2006). More recently, studies have demonstrated that tumour cells show a general down-regulation of miRNA expression compared with normal tissues. It has also been shown that the ratio of precursor miRNA to mature miRNA goes up in cancer, possibly pointing to a defect in processing, contributing to cancer cell phenotypes. An analysis of gene expression in primary tumours points to a failure in processing at the Drosha stage (Thomson et al. 2006). miRNAs have been shown to have a role in stem cell maintenance, differentiation and lineage determination (Wang et al. 2009b). This has been shown to be relevant in cancer by Yu et al., who demonstrated that breast cancer stem cells showed reduced levels of the miRNA let-7. In addition, reduced let-7 levels were required to maintain the undifferentiated state of these cells (Yu et al. 2007).

Different solid tumours express different miRNA expression profiles (Volinia et al. 2006; Yanaihara et al. 2006). A study, using bead-based flow cytometry, of miRNA expression in multiple human tumours, demonstrated that it had significantly greater diagnostic accuracy than a mRNA classification system (Lu et al. 2005). In fact, it is possible that miRNA classification could also detect pro-malignant states. The apparent richness of information that can be derived from miRNA expression profiles therefore has obvious diagnostic potential. It is possible to detect miRNA in the plasma, and this may prove to be an effective source of diagnostic markers of disease (Lawrie et al. 2008; Wang et al. 2009a).

miRNAs can exhibit behaviour as oncogenes or tumour suppressors. Let-7a-1 is located at chromosome 9q22.3, a locus frequently deleted in colon cancer. It was demonstrated that transfection of let-7a-1 precursors in to let-7 low-expressing DLD-1 colon cancer cells caused a reduction in growth rate, in a dose-dependent manner. Western blot analyses showed that expression of RAS and c-myc was reduced in the DLD-1 cells transfected with exogenous let-7, adding further evidence to its function as a tumour suppressor in this cell type (Akao et al. 2006). The miR-17-92 cluster consists of 6 miRNAs, and is located at 13q31, a locus which is amplified in a number of tumours. A study was performed using Eµ-myc transgenic mice (carrying the c-myc oncogene, driven by the immunoglobulin heavy chain enhancer Eµ), which develop B-cell lymphomas by 4–6 months of age. A truncated form of the miR-17 cluster, highly expressing all but one of the five miRNAs, was introduced into the haematopoietic cells of these mice. This resulted in a dramatic acceleration of disease onset and death (He et al. 2005).

4 miRNAs are Regulated by Hypoxia

Several studies have identified miRNAs which are regulated by hypoxia (Table 1)

Kulshreshtha et al. examined the effect of hypoxic stress in colon and breast cancer cell lines. Their study uncovered many hypoxia-regulated miRNAs (HRMs), though the majority were cell-line specific. However, a few HRMs were found to be consistently upregulated across the cell lines, including miR-21, 23a, 23b, 24, 26a, 26b, 27a, 30b, 93, 103, 103, 106a, 107, 125b, 181a, 181b, 181c, 192, 195, 210 and 213 (Kulshreshtha et al. 2007). Hua et al. identified miRNAs induced by hypoxia and desferrioxamine (DFOM) treatment in a human nasopharyngeal carcinoma epitheliod (CNE) cell line (Hua et al. 2006). The list of upregulated miRNAs differed substantially from that obtained by Kulshreshtha, probably due to the different cell lines, experimental conditions, and expression profiling technology employed. However, certain miRNAs, such as miR-210 and miR-181 were found to be upregulated in common. They also identified a group of miRNAs repressed by

Table 1 miRNAs up- and down-regulated in hypoxia

miRNAs up-regulated by hypoxia	Study	miRNAs down-regulated by hypoxia	Study
let-7b	5	Let-7-a	2
let-7e	5	Let-7-c	2
Let-7-i	3	Let-7-d	2
miR-103	1	Let-7-f	2
miR-106a	1	miR-101	3
miR-107	1	miR-122a	3
miR-125a	4, 5	miR-128b	4
miR-125b	1	miR-141	3
miR-128a	5	miR-15b	2
miR-137	5	miR-181d	4
miR-148a	3, 5	miR-186	3
miR-148b	3	miR-196a	4
miR-151	2	miR-196b	4
miR-152	4	miR-197	3
miR-15a	3	miR-19a	3
miR-181a	1	miR-200a*	4
miR-181c	1	miR-20a	2
miR-185	5	miR-20b	2
miR-188	2, 4	miR-216	5
miR-191	3, 4	miR-224	2
miR-192	1	miR-25	4
miR-193b	4	miR-29b	3
miR-199a	5	miR-30e-5p	3
miR-20	5	miR-320	3
miR-200a	3	miR-373*	4
miR-200b	4	miR-374	3
miR-204	5	miR-422b	3
miR-206	4	miR-424	3, 4
miR-21	1	miR-449	4
miR-210	1–5	miR-489	4
miR-213	1, 4, 5	miR-519e*	4
miR-214	3, 5	miR-565	3
miR-23a	1, 4	miR-9	5
miR-23b	1, 4, 5	miR-92	4
miR-24	1	miRNAs regulated by hypoxia with contrasting results between studies	Study and regulation (+ = up/ − = down)
miR-26a	1, 4		
miR-27a	1		
miR-27b	4		
miR-299	5		
miR-30a-3p	5	Let-7-e	3(+); 2(−)
miR-30a-5p	4	Let-7-g	3(+); 2(−)
miR-30c	4, 5	miR-150	5(+); 4(−)
miR-30d	2, 4	miR-155	2, 5(+); 4(−)
miR-335	5	miR-16	5(+); 2(−)
miR-339	4	miR-181b	1, 2, 5(+); 4(−)
miR-342	5	miR-195	1 (+); 3(−)
miR-373	3	miR-26b	1, 5(+); 2(−)
miR-429	3	miR-30b	1, 3, 5(+); 2(−)
miR-452*	4		
miR-491	4		

(continued)

Table 1 (continued)

miRNAs up-regulated by hypoxia	Study	miRNAs down-regulated by hypoxia	Study
miR-498	3		
miR-512-5p	4		
miR-563	3		
miR-572	3		
miR-628	3		
miR-637	3		
miR-7	3		
miR-93	1, 4		
miR-98	3		

Studies: *1* colon and breast cancer cells, 0.2% O_2, 8–48 h (Kulshreshtha et al. 2007); *2* nasopharyngeal carcinoma cells, DFOM treatment, 20 h (Hua et al. 2006); *3* head and neck squamous carcinoma cells, 1% O_2, 1 h or 5% O_2, 8 h (Hebert et al. 2007); *4* primary human cytotrophoblasts, 1% O_2, 48 h (Donker et al. 2007); *5* colon cells, liquid–liquid interface (Guimbellot et al. 2009); *6* HUVEC Cells, 1% O_2, 24 h (Pulkkinen et al. 2008)

hypoxia. Four of these miR-15, 16, 20a and 20b, were found to translationally regulate Vascular Endothelial Growth Factor (VEGF) in luciferase reporter assays, providing evidence for direct miRNA-mediated regulation of a key angiogenic pathway (Hua et al. 2006). Of note, miR-15 and 16 are also known to have tumour suppressor activity in chronic lymphocytic leukaemia through targeting the anti-apoptotic protein BCL2. miR-15 and 16 are frequently deleted or downregulated in these tumours (Cimmino et al. 2005).

Hypoxic regulation of miRNAs does not appear to be due to changes in the processing machinery of these molecules. Donker et al. showed that miRNAs were up- or downregulated in hypoxia trophoblast cells, but with no change in the expression of Drosha, Exportin 5, Dicer, Ago2 and DP103 (DEAD box protein 103) at mRNA level (Donker et al. 2007). miRNAs are frequently regulated by transcription factors induced by stress pathways, which are also regulated by hypoxia independently of HIF (Hypoxia Inducible Factor). For example, the miR-34 family is induced by DNA damage and in a p53-dependent manner (He et al. 2007), and miR-155 is transcriptionally induced by protein kinase C (PKC) and NF-κB (nuclear factor kappa-light-chain-enhancer of activated B cells). Both p53 and NF-κB have been implicated in the response of tumours to hypoxia (Kluiver et al. 2007).

Hypoxia regulated miRNAs from cell line studies (Table 1) showed a variety of patterns in primary breast cancer (Fig. 2). However, two main patterns of miRNAs expression could be distinguished, one set correlating with the hypoxia-regulated miR-210, and a contrary pattern. Based on the expression of these miRNAs, breast cancer patient samples could be stratified into two defined clusters (Fig. 2). Cluster 1 was characterised by low expression of miRNAs upregulated by hypoxia, such as miR-210, -24, -27a, whereas these miRNAs were over-expressed in cluster 2. Furthermore, cluster 1 showed a significantly lower (Mann–Whitney test, $Z = -4.9$, $p < 0.001$) expression of a mRNA based hypoxia-score derived from a common head and neck and breast cancer hypoxia signature (Buffa et al. 2010).

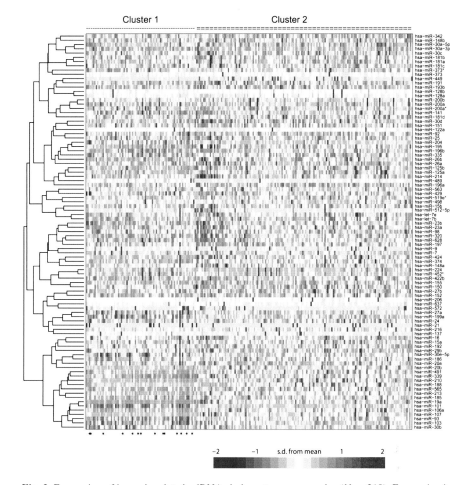

Fig. 2 Expression of hypoxia related miRNAs in breast cancer samples ($N = 219$). Expression is measured by Illumina miRNA arrays (Illumina Inc., San Diego), normalised and logged (base2), and centred and standardised by gene (see *bar*). *X*-axis: 93 hypoxia-related miRNAs from cell lines studies (Table 1) clustered on their expression pattern using hierarchical clustering with Pearson correlation. *Y*-axis: breast cancer samples and normal tissue pools (indicated with a *asterisk*) clustered in two groups using BIRCH, a recursive hierarchical clustering algorithm with Bayesian optimisation of the number of clusters (Chiu et al. 2001). This method allows objective optimisation of the number of clusters using a Bayesian Information Criterion

5 Hypoxia Inducible Factor and miR-210

HIFs are heterodimeric transcription factors consisting of a HIF-α molecule bound to an aryl receptor nuclear translocator (ARNT, or HIF-β). In normoxia, the HIF-α molecules, which exist as HIF-1,-2 or -3α, undergo prolyl hydroxylation and bind to the von-Hippel Lindau protein which targets them for proteaosomal degradation through ubiquitination. In hypoxia, hydroxylation does not occur and the HIF-α molecules are stabilised and are able to bind to the constitutively expressed ARNT (Harris 2002). Several studies now point to miR-210 as the miRNA most ubiquitously upregulated by HIF across different tumour types and cell lines. It is upregulated in the placentas of women with pre-eclampsia, a condition in which inadequate blood supply to the placenta is known to cause tissue hypoxia (Pineles et al. 2007; Zhu et al. 2009). It has been confirmed as a HIF-1 target through knockout of HIFs using siRNA, constitutive over-expression studies, and chromatin immunoprecipitation (Kulshreshtha et al. 2007; Camps et al. 2008).

6   HRMs and Cancer Prognosis

miRNAs have long been shown to be associated with cancer, but more recently, studies have provided links of HRMs with clinical outcome. miR-210 is associated with triple-negative lymph-node negative breast cancer (Foekens et al. 2008). Greither et al. found that miR-210 and 155, both HRMs, were associated with poor outcome in pancreatic adenocarcinoma, along with two other miRNAs, miR-203 and miR-222 (Greither et al. 2010).

High miR-210 expression has been shown to be correlated with a poor outcome in breast cancer, measured as both disease-free and overall survival. This was true both when miR-210 expression was measured by qPCR (Camps et al. 2008) and by Illumina microarrays (Fig. 3a) in a cohort of 210 early invasive breast cancers. Work from our group (Buffa et al. in submission) found that the expression of the transcript

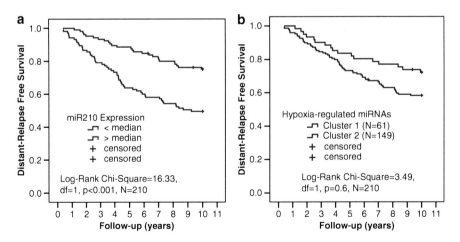

Fig. 3 Association of miR-210 and a hypoxia miRNA signature with Distant Relapse Free Survival. Log-rank test for the null hypothesis that the groups have equal survival is shown. (**a**) There is no clinically defined cut-off point for miR-210 over-expression, thus patients were stratified into two equal-size groups by miR-210 median value. (**b**) Patients were grouped using the two clusters shown in Fig. 2

containing the miR-210 precursor is also prognostic in independent breast cancer datasets (Spearman rho $= 0.45$, $p < 0.001$, $N = 216$). In addition, miR-210 expression in breast cancer correlated positively with tumour hypoxia, determined through a hypoxia-score based on the expression profiling of a 99-gene hypoxia metagene (Camps et al. 2008) and a common head and neck and breast cancer hypoxia signature (Spearman rho $= 0.45$, $p < 0.001$, $N = 216$) (Buffa et al. in submission). We also found this in head and neck squamous cell carcinoma (Gee et al. 2010).

A primary tumour signature of hypoxia-related miRNAs from cell line studies (Fig. 2) showed a prognostic trend in breast cancer (Fig. 3b) although this was not significant ($p = 0.06$). However, a signature of hypoxia-related miRNAs derived by direct data-mining of breast cancer data was shown to be a significant independent prognostic factor in breast cancer in multivariate analysis correcting for clinico-pathological variables (Buffa et al. in submission).

7 HRM Targets

There has been considerable interest in miR-210 due to its consistent association with hypoxia in both cell lines and tissues, and recently several publications have uncovered some of its biological targets (Table 2).

Fasanaro et al. found that miR-210 was necessary for the formation of capillary-like structures on Matrigel, in the U2OS osteosarcoma cell line. They also demonstrated that miR-210 increased cell migration in response to VEGF. They showed that EphrinA3 (EFNA3) was targeted and downregulated at protein level in HUVECs by miR-210, using immunofluorescence and luciferase reporter assays. Down-regulation of EFNA3 by miR-210 was necessary for tubulogenesis (Fasanaro

Table 2 Confirmed gene targets of miR-210

Gene symbol	Gene name	Study
EFNA3	Ephrin-A3	1, 2
RAD52	RAD52 homolog (*S. cervisiae*)	3
MNT	MAX-binding protein	4
HOXA1	Homeobox A1	5
HOXA9	Homeobox A9	5
FGFRL1	Fibroblast growth factor-like 1	5
ISCU	Iron sulphur cluster scaffold homolog	6, 7
CASP8P	Caspase8-associated protein 2	8
NPTX1	Neuronal pentraxin 1	2
ACVR1B	Activin receptor 1B	9
E2F3	E2F transcription factor 3	10
BDNF	Brain-derived neurotrophic factor	11
PTPN1	Tyrosine-protein phosphatase non-receptor type 1	11
P4HB	Protein disulphide isomerase	11
GPD1L	Glycerol-3-phosphate dehydrogenase 1-like	11

Studies: *1* Fasanaro et al. (2008), *2* Pulkkinen et al. (2008); *3* Crosby et al. (2009); *4* Zhang et al. (2009); *5* Huang et al. (2009); *6* Tong and Rouault (2006); *7* Chan et al. (2009); *8* Kim et al. (2009); *9* Mizuno et al. (2009); *10* Giannakakis et al. (2008); *11* Fasanaro et al. (2009)

et al. 2008). The Ephrins are receptor tyrosine protein kinases which have been implicated in both development and erythropoeisis. The above work suggests a mechanism by which miRNA is linked to angiogenesis.

Crosby et al. showed that miR-210 and miR-373, both upregulated in hypoxia in HeLa (cervical cancer) and MCF7 (breast cancer) cell lines, target the protein RAD52. RAD52 is a key factor involved in DNA double-strand break repair and homologous recombination. RAD52 was downregulated in hypoxia, and this effect was partially reversed by antisense inhibition of miR-210. miRNA mediation of the DNA repair pathways may therefore be an important factor in genetic instability in these tumour types (Crosby et al. 2009).

Zhang et al. showed that miR-210 derepresses c-myc function in hypoxia by targeting MNT, a transcription factor which is known to antagonise myc function. As HIF-1, but not HIF-2 is known to also antagonise myc, expression of miR-210 may serve to balance the effects of HIF-1 and therefore exert finer control over myc function (Foshay and Gallicano 2007; Zhang et al. 2009).

Bianchi et al. showed an association of enhanced miR-210 levels with erythroid differentiation and production of HbF (foetal haemoglobin), although they offered no direct evidence of the role of the miRNA in this process (Bianchi et al. 2009).

Elevated VEGF levels in B-cell chronic lymphocytic leukaemia are associated with advancing disease. These cells are able to spontaneously secrete VEGF, and its expression has been linked to apoptosis resistance. It has been established that B-CLL cells over-express HIF-1α in normoxia, in the presence of reduced expression of VHL. Ghosh et al. found that miR-92-1 is over-expressed in B-CLL, and targets VHL protein, providing a mechanism for increased HIF-1α activity in this cancer type (Ghosh et al. 2009).

Huang et al. investigated targets of miR-210 by using microarrays to identify mRNAs which immunoprecipitated with Ago2, the primary component of the miRNP complex. They validated Homeobox A1 (HOXA1), HOXA9 and Fibroblast Growth Factor-like 1 (FGFRL1), with luciferase reporter constructs. The 3′-UTR of FGFRL1 contained seven potential binding sites of miR-210, and was thus selected for xenograft studies along with HOXA1. They found that over-expression of miR-210 in head and neck and pancreatic cancer cell xenografts in mice inhibited the initial phase of tumour growth. They observed a partial rescue with over-expression of FGFRL1 or HOXA1, suggesting that multiple genes are involved in the inhibition of tumour growth initiation (Huang et al. 2009). Fasanaro et al. identified a number of targets by immunoprecipitation of miR-210 enriched RISC, and also through 2D gel and proteomic studies. They found that some proven and/or predicted miR-210 targets, such as EFNA3, E2F3 (Giannakakis et al. 2008), NPTX1 (Pulkkinen et al. 2008), RAD52 and ACVR1B (Mizuno et al. 2009) were enriched in the immunoprecipitate. They confirmed new candidate miR-210 targets such as brain-derived neurotrophic factor (BDNF), tyrosine-protein phosphatase non-receptor type 1 (PTPN1), protein disulphide isomerase (P4HB), and glycerol-3-phosphate dehydrogenase 1-like (GPD1L). Of note, they found that in some of the targets, the miR-210 seed pairing region was located in the protein-coding or 5′-UTR region of the target mRNA, supporting other recent evidence that miRNA-binding is not restricted to the classic 3'-UTR model (Orom et al. 2008; Fasanaro et al. 2009; Tsai et al. 2009).

7.1 miR210 Regulation of ISCU and Mitochondrial Metabolism

miR-210 has recently been shown to target and downregulate the iron sulphur cluster protein (ISCU). This protein acts as a scaffold for the assembly of iron sulphur clusters [Fe–S], which are critical cofactors for enzymes involved in electron transport, the Krebs cycle and iron metabolism. The Fe–S cluster proteins are key components of the mitochondrial electron transport chain complexes I, II and III, in addition to enzymes of the Krebs cycle: aconitase and the succinate dehydrogenase activity of complex II (Rouault and Tong 2008).

Using mice carrying tamoxifen-dependent conditionally silenced VHL alleles, Chan et al. demonstrated that miR-210 is present in much higher levels in the absence of VHL in the kidney, liver and heart. Correspondingly, there was a reciprocal relationship between VHL expression and ISCU protein levels in these organs. In vitro assays demonstrated that inhibition of ISCU by miR-210, and thus inhibition of iron–sulphur cluster formation, caused a reduction in aconitase and mitochondrial complex I activity in hypoxia (Chan et al. 2009).

Work from our group concentrated on cancer cells, rather than primary cells. We found that ISCU protein and mRNA in MCF7 and HCT116 (breast cancer) cell lines were downregulated in response to hypoxia alone or exogenous miR-210 transfection. In addition to the effects of miR-210 on aconitase and complex I activity, we found a striking effect of miR-210 on free radical production. Addition of miR-210 in normoxia significantly increased superoxide production at 48 h, as measured by MitoSox staining. Exposure of cells to hypoxia also increased super-oxide, as expected, and this effect was completely reversed with transfection of miR-210 inhibitors. Cotransfection of a plasmid containing the ISCU2 coding sequence rescued the induction of superoxide production by miR-210 in normoxia. Increased miR-210 in normoxia led to an increase in glycolysis, characterised by an increase in lactate production and a reduction in pyruvate. We found that miR-210 inhibitors led to decreased lactate production in hypoxia.

We found that another consequence of ISCU down-regulation by miR-210 is uptake of iron by the HCT116 cells. This is likely mediated by cytoplasmic aconitase which, upon depletion of its Fe–S cluster, acts a translational regulator, and is known as iron-responsive element (IRE)-binding protein 1 (IRP1). Increased IRE binding leads to the post transcriptional repression of ferritin synthesis, and increased expression of transferrin receptor, leading to increased cellular uptake of iron (Tong and Rouault 2006) (Fig. 4).

Finally, we found that ISCU expression was significant in vivo. Xenografts of the glioblastoma cell line U87, which had been treated with the VEGF inhibitor Avastin (Bevacizumab) demonstrated Avastin-induced necrosis, and increased expression of HIF-1α and its target genes, carbonic anhydrase 9 (CA9) and VEGF.

cytosolic iron–sulfur protein assembly CIA iron–sulfur cluster ISC

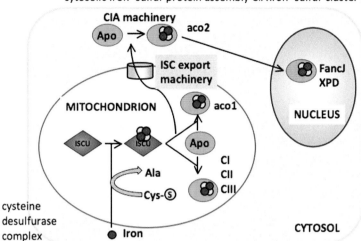

Fig. 4 The role of iron sulphur protein cluster homologue ISCU in metabolism. Iron is transported into the mitochondria by a specific transport system and metabolised by a cysteine desulphurase complex to form iron sulphur complexes, which are bound and chaperoned by ISCU. These iron sulphur complexes are critical for active sites of many enzymes, which are synthesised as apoenzymes. This applies to components of complex 1, 2 and 3 of the electron transport chain and ISCU is critical in integrating the iron sulphur complexes into these subunits, to allow effective ATP generation. The iron sulphur complexes are also exported by a specific transport pathway to be used to assemble with apoproteins in the cytosol, with a separate set of chaperones involved in cytostolic iron sulphur protein assembly. Amongst these enzymes is Aconitase 2 and also nuclear enzymes involved in DNA repair such as FANCJ and XPD. Thus, ISCU down-regulation will affect many fundamental metabolism processes

These tumours were found to have marked up-regulation of miR-210, and reciprocal down-regulation of ISCU mRNA. Analysis of our breast (213 patients) and head and neck (43 patients) tumour series showed a highly significant inverse relationship of miR-210 to ISCU expression. ISCU expression in these series was significantly and negatively correlated with patient prognosis.

Thus, overall effects of a single miRNA are complex with both pro and anti-tumorogenic effects. This will be further affected by gene expression profiles of target genes and degree of induction which varies many folds between cell lines. However, the strong clinical data suggests that the overall effect is positive for tumour growth or metastasis.

7.2 HRMs and Apoptosis

A mechanism by which HIF-1α mediates hypoxia-induced apoptosis is by stabilising p53 (An et al. 1998). Overexpression of miR-199a abolishes the induction of HIF-1α and p53 in myocytes in the first 24 h of exposure to hypoxia. Rane et al. found that miR-199a is rapidly reduced to undetectable levels during hypoxia. Hypoxia did not affect miR-199a*, which is expressed from the same stem loop, nor did it affect miR-1, suggesting that it is a specific effect. In addition, levels of the miR-199a precursor continued to accumulate during hypoxia, suggesting that the effect is post-transcriptionally mediated. They found that Sirtuin-1 induced down-regulation of prolyl hydroxylase 2 (PHD2) is required for HIF-1α accumulation in myocytes during hypoxic pre-conditioning. In addition, HIF-1α appears to be a direct target of this miRNA. They found that down-regulation of miR-199a during hypoxia was necessary for induction of caspases −3, −6, −9, −12, and FasL, AIF and BNIP1. Over-expression of miR-199a completely abolished the hypoxia-induced expression of these genes (Rane et al. 2009).

Taguchi et al. identified HIF-1α as a target for the miR-17–92 cluster. They found that the cluster regulated HIF expression in normoxia. However, expression of the miR-17–92 cluster did not change in hypoxia, nor did it regulate HIF in hypoxic conditions in their ACC-LC-172 and Calu6 lung cancer cell lines (Taguchi et al. 2008). Interestingly, Yan et al. found that hypoxia *did* induce down-regulation of members of the miR-17–92 cluster (miRs 18a, 19a, 19b and 20a) in hypoxia in HCT116 p53$^{+/-}$) Caco-2 cells. They confirmed the role of p53 by comparing HCT116 p53$^{+/-}$; as expected, the absence of p53 abolished hypoxic repression of the miR-17–92 cluster. Repression of c-myc by siRNA reduced normoxic levels of the miR-17–92 cluster, but did not inhibit further reduction in expression in hypoxia. Conversely, p53 siRNA had little effect on normoxic miR-17–92 levels, but abolished its down-regulation in hypoxia. They showed that p53 has cis-regulator HRE activity, and that it competes with TATA-binding protein (TBP) for a site in the miR-17–92 promotor region in hypoxic conditions (Yan et al. 2009). As previously discussed, the miR-17–92 cluster targets the anti-apoptotic BCL2. Down-regulation of this cluster by p53 may therefore represent part of the mechanism for hypoxia-induced apoptosis (Xiao et al. 2008). This example demonstrates the complex regulation of miRNAs in hypoxia, and goes some way toward explaining the huge variability of individual cell-type responses to stimuli such as hypoxia.

7.3 Ischaemic Preconditioning

Ischaemic preconditioning (IP) is a powerful cytoprotective stimulus for stem cells. Cells subjected to cycles of hypoxia and reoxygenation, have improved survival in more prolonged hypoxic conditions (Kim et al. 2009). It has recently been shown that transplantation of such pre-conditioned cells in to infarcted heart improves its function via enhanced survival of implanted cells and angiogenesis (Hu et al. 2008). It is currently widely accepted that hypoxia in the tumour microenvironment is typically cyclical, with fluctuating degrees of oxygenation (Dewhirst et al. 2008).

The mechanism by which IP increases stem-cell survival and angiogenesis is, therefore, likely to be equally important in the clonogenic stem cell population in cancer. Kim et al. showed that miR-210 is upregulated in mesenchymal stem cells (MSCs) following ischaemic pre-conditioning. They demonstrated that miR-210 targets and reduces expression of caspase-8-associated protein 2 (CASP8AP2) twofold, following IP in MSCs exposed to anoxia. miR-210 knockdown in IP cells resulted in increased CASP8AP2 expression and increased cell death. Additionally, knock-down of CASP8AP2 in non-preconditioned cells improved their survival under anoxia (Kim et al. 2009). While stem cell survival has a clear advantage in a transplantation environment, one would predict that increased survival in tumour cells secondary to reduced CASP8APS would adversely affect prognosis. Indeed, it has been shown that CASP8AP2 deletions in T-cell acute lymphoblastic leukaemia correlate with poor early treatment response (Remke et al. 2009), and similar findings were made in acute lymphoblastic leukaemia (Flotho et al. 2006).

8 Target Prediction

Accurate target prediction remains problematic and is one of the major factors holding back research into miRNA function. Several bioinformatics methods and algorithms have been suggested for target prediction (Barbato et al. 2009). In general, these identify targets by searching for potential binding sites in the 3′-UTR and other transcript regions. However, results from these algorithms present large discrepancies, and accurate prediction of miRNA targets is still a challenging goal. These discrepancies are partially due to differences in the algorithms and their implementation, and to different requirements for site conservation across species. They are also due to the fact that there are different hypotheses on the miRNA action on its target genes. For example, a recent study showed that targets containing simultaneous 5′- and 3′-UTR miRNA interaction sites can identify targets with higher expression modulation (Lee et al. 2009).

Due to these differences, considering predictions from multiple methods might be more effective than focussing on one method, or considering only the overlapping predictions from the different methods (Ritchie et al. 2010). In this respect, work from our group (Buffa et al. in submission) employed a simple non-parametric ranked score combining predictions from different algorithms. All predictions from the considered methods are regarded as potential targets, but greater weight can be assigned to targets predicted by more than one algorithm. Such a method allows inclusion of other criteria for target selection; for example miRNA action at the mRNA and protein level can be used as one of the criteria to select miRNA targets. This has been shown in a recent study of miR-210 targets by Fasanaro et al. where combined analysis of mRNA and protein expression changes after modulation of miR-210 proved a good method to identify miR-210 targets (Fasanaro et al. 2009).

Furthermore, combining miRNA expression and mRNA expression information can be used to select potential targets using material from retrospective clinical series (Buffa et al. in submission). Specifically, potential targets that have mRNA levels inversely correlated to miRNA expression levels, are considered as having the correct mRNA expression behaviour to be real targets. This method has been recently applied to a large breast cancer dataset ($N = 219$) and has identified the iron–sulphur cluster scaffold homolog (ISCU, alias NIFU) as a miR-210 target (Favaro et al. 2010), which was confirmed by experimental validation both in cancer and normal tissues (Chan et al. 2009).

9 Future Clinical Applications of HRM Research

It is becoming clear that the roles of HRMs span a number of processes which are relevant to both metabolic adaptation to hypoxia and oncogenesis. HRMs have been shown to be involved in regulation of the well-described oncogene myc, in addition to being implicated in the control of apoptosis, angiogenesis, erythroid differentiation, DNA repair and mitochondrial metabolism. It is clear that differential regulation of miRNAs in hypoxia varies significantly between cell types, but one, miR-210, stands out as the most ubiquitously upregulated miRNA across both primary and tumour cell lines. This miRNA is upregulated in vivo and correlates with poor prognosis in patients, in addition to correlating strongly with hypoxic gene signatures. Our ability to detect miRNA in plasma, urine and tumour tissues shows the potential of these molecules to act as predictors of survival or response to oncological treatments.

It has already been demonstrated that injection of miRNA antagonists (antago-Mirs) can deliver a therapeutic effect. For example, Esau et al. showed that intravenous administration of anti-miR-122 in mice resulted in reduced plasma cholesterol levels and increased fatty acid oxidation (Esau et al. 2006). As we learn more about the biological functions of HRMs, these miRNAs may ultimately prove useful as therapeutic targets themselves, alone or in conjunction with other agents.

References

Buffa FM, Camps C, Winchester L, Gee H, Sheldon H, Taylor M, Harris AL, Ragoussis J Integration of miRNA and mRNA expression profiles in breast cancer identifies prognostic markers and associated pathways (in submission)

Akao Y, Nakagawa Y, Naoe T (2006) let-7 microRNA functions as a potential growth suppressor in human colon cancer cells. Biol Pharm Bull 29:903–906

An WG, Kanekal M, Simon MC, Maltepe E, Blagosklonny MV, Neckers LM (1998) Stabilization of wild-type p53 by hypoxia-inducible factor 1alpha. Nature 392:405–408

Anderson P, Kedersha N (2006) RNA granules. J Cell Biol 172:803–808

Bagga S, Bracht J, Hunter S, Massirer K, Holtz J, Eachus R, Pasquinelli AE (2005) Regulation by let-7 and lin-4 miRNAs results in target mRNA degradation. Cell 122:553–563

Barbato C, Arisi I, Frizzo ME, Brandi R, Da Sacco L, Masotti A (2009) Computational challenges in miRNA target predictions: to be or not to be a true target? J Biomed Biotechnol 2009:803069

Bartel DP, Chen CZ (2004) Micromanagers of gene expression: the potentially widespread influence of metazoan microRNAs. Nat Rev Genet 5:396–400

Bianchi N, Zuccato C, Lampronti I, Borgatti M, Gambari R (2009) Expression of miR-210 during erythroid differentiation and induction of gamma-globin gene expression. BMB Rep 42:493–499

Brennecke J, Stark A, Russell RB, Cohen SM (2005) Principles of microRNA-target recognition. PLoS Biol 3:e85

Buffa FM, Harris AL, West CM, Miller CJ (2010) Large meta-analysis of multiple cancers reveals a common, compact and highly prognostic hypoxia metagene. Br J Cancer 102:428–435

Camps C, Buffa FM, Colella S, Moore J, Sotiriou C, Sheldon H, Harris AL, Gleadle JM, Ragoussis J (2008) hsa-miR-210 Is induced by hypoxia and is an independent prognostic factor in breast cancer. Clin Cancer Res 14:1340–1348

Chan SY, Zhang YY, Hemann C, Mahoney CE, Zweier JL, Loscalzo J (2009) MicroRNA-210 controls mitochondrial metabolism during hypoxia by repressing the iron-sulfur cluster assembly proteins ISCU1/2. Cell Metab 10:273–284

Chiu AG, Newkirk KA, Davidson BJ, Burningham AR, Krowiak EJ, Deeb ZE (2001) Angiotensin-converting enzyme inhibitor-induced angioedema: a multicenter review and an algorithm for airway management. Ann Otol Rhinol Laryngol 110:834–840

Chu CY, Rana TM (2006) Translation repression in human cells by microRNA-induced gene silencing requires RCK/p54. PLoS Biol 4:e210

Cimmino A, Calin GA, Fabbri M, Iorio MV, Ferracin M, Shimizu M, Wojcik SE, Aqeilan RI, Zupo S, Dono M, Rassenti L, Alder H, Volinia S, Liu CG, Kipps TJ, Negrini M, Croce CM (2005) miR-15 and miR-16 induce apoptosis by targeting BCL2. Proc Natl Acad Sci USA 102:13944–13949

Crosby ME, Kulshreshtha R, Ivan M, Glazer PM (2009) MicroRNA regulation of DNA repair gene expression in hypoxic stress. Cancer Res 69:1221–1229

Dewhirst MW, Cao Y, Moeller B (2008) Cycling hypoxia and free radicals regulate angiogenesis and radiotherapy response. Nat Rev Cancer 8:425–437

Diederichs S, Haber DA (2007) Dual role for argonautes in microRNA processing and posttranscriptional regulation of microRNA expression. Cell 131:1097–1108

Doench JG, Sharp PA (2004) Specificity of microRNA target selection in translational repression. Genes Dev 18:504–511

Donker RB, Mouillet JF, Nelson DM, Sadovsky Y (2007) The expression of Argonaute2 and related microRNA biogenesis proteins in normal and hypoxic trophoblasts. Mol Hum Reprod 13:273–279

Esau C, Davis S, Murray SF, Yu XX, Pandey SK, Pear M, Watts L, Booten SL, Graham M, McKay R, Subramaniam A, Propp S, Lollo BA, Freier S, Bennett CF, Bhanot S, Monia BP (2006) miR-122 regulation of lipid metabolism revealed by in vivo antisense targeting. Cell Metab 3:87–98

Eulalio A, Behm-Ansmant I, Izaurralde E (2007) P bodies: at the crossroads of post-transcriptional pathways. Nat Rev Mol Cell Biol 8:9–22

Fasanaro P, D'Alessandra Y, Di Stefano V, Melchionna R, Romani S, Pompilio G, Capogrossi MC, Martelli F (2008) MicroRNA-210 modulates endothelial cell response to hypoxia and inhibits the receptor tyrosine kinase ligand Ephrin-A3. J Biol Chem 283:15878–15883

Fasanaro P, Greco S, Lorenzi M, Pescatori M, Brioschi M, Kulshreshtha R, Banfi C, Stubbs A, Calin GA, Ivan M, Capogrossi MC, Martelli F (2009) An integrated approach for experimental target identification of hypoxia-induced miR-210. J Biol Chem 284:35134–35143

Favaro E, Ramachandran A, McCormick R, Gee H, Blancher C, Crosby M, Devlin C, Blick C, Buffa F, Li JL, Neves R, Glazer P, Iborra F, Ivan M, Ragoussis J, Harris AL (2010) MicroRNA-210 regulates mitochondrial free radical response to hypoxia and krebs cycle in cancer cells by cluster protein ISCU. PLoS One 5:e10345

Filipowicz W, Bhattacharyya SN, Sonenberg N (2008) Mechanisms of post-transcriptional regulation by microRNAs: are the answers in sight? Nat Rev Genet 9:102–114

Flotho C, Coustan-Smith E, Pei D, Iwamoto S, Song G, Cheng C, Pui CH, Downing JR, Campana D (2006) Genes contributing to minimal residual disease in childhood acute lymphoblastic leukemia: prognostic significance of CASP8AP2. Blood 108:1050–1057

Foekens JA, Sieuwerts AM, Smid M, Look MP, de Weerd V, Boersma AW, Klijn JG, Wiemer EA, Martens JW (2008) Four miRNAs associated with aggressiveness of lymph node-negative, estrogen receptor-positive human breast cancer. Proc Natl Acad Sci USA 105:13021–13026

Foshay KM, Gallicano GI (2007) Small RNAs, big potential: the role of MicroRNAs in stem cell function. Curr Stem Cell Res Ther 2:264–271

Gee HE, Camps C, Buffa FM, Patiar S, Winter SC, Betts G, Homer J, Corbridge R, Cox G, West CM, Raoussis J, Harris AL (2010) hsa-mir-210 is a marker of tumor hypoxia and a prognostic factor in head and neck cancer. Cancer 116:2148–2158

Ghosh AK, Shanafelt TD, Cimmino A, Taccioli C, Volinia S, Liu CG, Calin GA, Croce CM, Chan DA, Giaccia AJ, Secreto C, Wellik LE, Lee YK, Mukhopadhyay D, Kay NE (2009) Aberrant regulation of pVHL levels by microRNA promotes the HIF/VEGF axis in CLL B cells. Blood 113:5568–5574

Giannakakis A, Sandaltzopoulos R, Greshock J, Liang S, Huang J, Hasegawa K, Li C, O'Brien-Jenkins A, Katsaros D, Weber BL, Simon C, Coukos G, Zhang L (2008) miR-210 links hypoxia with cell cycle regulation and is deleted in human epithelial ovarian cancer. Cancer Biol Ther 7:255–264

Greither T, Grochola F, Udelnow A, Lautenschlager C, Wurl P, Taubert H (2010) Elevated expression of microRNAs 155, 203, 210 and 222 in pancreatic tumours associates with poorer survival. Int J Cancer 126:73–80

Grimson A, Farh KK, Johnston WK, Garrett-Engele P, Lim LP, Bartel DP (2007) MicroRNA targeting specificity in mammals: determinants beyond seed pairing. Mol Cell 27:91–105

Guimbellot JS, Erickson SW, Mehta T, Wen H, Page GP, Sorscher EJ, Hong JS (2009) Correlation of microRNA levels during hypoxia with predicted target mRNAs through genome-wide microarray analysis. BMC Med Genomics 2:15

Harris AL (2002) Hypoxia–a key regulatory factor in tumour growth. Nat Rev Cancer 2:38–47

Hayashita Y, Osada H, Tatematsu Y, Yamada H, Yanagisawa K, Tomida S, Yatabe Y, Kawahara K, Sekido Y, Takahashi T (2005) A polycistronic microRNA cluster, miR-17-92, is overexpressed in human lung cancers and enhances cell proliferation. Cancer Res 65:9628–9632

He L, Thomson JM, Hemann MT, Hernando-Monge E, Mu D, Goodson S, Powers S, Cordon-Cardo C, Lowe SW, Hannon GJ, Hammond SM (2005) A microRNA polycistron as a potential human oncogene. Nature 435:828–833

He X, He L, Hannon GJ (2007) The guardian's little helper: microRNAs in the p53 tumor suppressor network. Cancer Res 67:11099–11101

Hebert C, Norris K, Scheper MA, Nikitakis N, Sauk JJ (2007) High mobility group A2 is a target for miRNA-98 in head and neck squamous cell carcinoma. Mol Cancer 6:5

Hu X, Yu SP, Fraser JL, Lu Z, Ogle ME, Wang JA, Wei L (2008) Transplantation of hypoxia-preconditioned mesenchymal stem cells improves infarcted heart function via enhanced survival of implanted cells and angiogenesis. J Thorac Cardiovasc Surg 135:799–808

Hua Z, Lv Q, Ye W, Wong CK, Cai G, Gu D, Ji Y, Zhao C, Wang J, Yang BB, Zhang Y (2006) MiRNA-directed regulation of VEGF and other angiogenic factors under hypoxia. PLoS One 1:e116

Huang X, Ding L, Bennewith KL, Tong RT, Welford SM, Ang KK, Story M, Le QT, Giaccia AJ (2009) Hypoxia-inducible mir-210 regulates normoxic gene expression involved in tumor initiation. Mol Cell 35:856–867

Hwang HW, Mendell JT (2006) MicroRNAs in cell proliferation, cell death, and tumorigenesis. Br J Cancer 94:776–780

Jakymiw A, Lian S, Eystathioy T, Li S, Satoh M, Hamel JC, Fritzler MJ, Chan EK (2005) Disruption of GW bodies impairs mammalian RNA interference. Nat Cell Biol 7:1267–1274

Kapp LD, Lorsch JR (2004) The molecular mechanics of eukaryotic translation. Annu Rev Biochem 73:657–704

Kim HW, Haider HK, Jiang S, Ashraf M (2009) Ischemic preconditioning augments survival of stem cells via MIR-210 expression by targeting caspase-8 associated protein 2. J Biol Chem 284(48):33161–33168

Kinch LN, Grishin NV (2009) The human Ago2 MC region does not contain an eIF4E-like mRNA cap binding motif. Biol Direct 4:2

Kiriakidou M, Tan GS, Lamprinaki S, De Planell-Saguer M, Nelson PT, Mourelatos Z (2007) An mRNA m7G cap binding-like motif within human Ago2 represses translation. Cell 129:1141–1151

Kluiver J, van den Berg A, de Jong D, Blokzijl T, Harms G, Bouwman E, Jacobs S, Poppema S, Kroesen BJ (2007) Regulation of pri-microRNA BIC transcription and processing in Burkitt lymphoma. Oncogene 26:3769–3776

Kulshreshtha R, Ferracin M, Negrini M, Calin GA, Davuluri RV, Ivan M (2007) Regulation of microRNA expression: the hypoxic component. Cell Cycle 6:1426–1431

Lawrie CH, Gal S, Dunlop HM, Pushkaran B, Liggins AP, Pulford K, Banham AH, Pezzella F, Boultwood J, Wainscoat JS, Hatton CS, Harris AL (2008) Detection of elevated levels of tumour-associated microRNAs in serum of patients with diffuse large B-cell lymphoma. Br J Haematol 141:672–675

Lee RC, Feinbaum RL, Ambros V (1993) The C. elegans heterochronic gene lin-4 encodes small RNAs with antisense complementarity to lin-14. Cell 75:843–854

Lee I, Ajay SS, Yook JI, Kim HS, Hong SH, Kim NH, Dhanasekaran SM, Chinnaiyan AM, Athey BD (2009) New class of microRNA targets containing simultaneous 5′-UTR and 3′-UTR interaction sites. Genome Res 19:1175–1183

Lim LP, Lau NC, Garrett-Engele P, Grimson A, Schelter JM, Castle J, Bartel DP, Linsley PS, Johnson JM (2005) Microarray analysis shows that some microRNAs downregulate large numbers of target mRNAs. Nature 433:769–773

Liu J, Rivas FV, Wohlschlegel J, Yates JR 3rd, Parker R, Hannon GJ (2005) A role for the P-body component GW182 in microRNA function. Nat Cell Biol 7:1261–1266

Lu J, Getz G, Miska EA, Alvarez-Saavedra E, Lamb J, Peck D, Sweet-Cordero A, Ebert BL, Mak RH, Ferrando AA, Downing JR, Jacks T, Horvitz HR, Golub TR (2005) MicroRNA expression profiles classify human cancers. Nature 435:834–838

Meister G (2007) miRNAs get an early start on translational silencing. Cell 131:25–28

Mizuno Y, Tokuzawa Y, Ninomiya Y, Yagi K, Yatsuka-Kanesaki Y, Suda T, Fukuda T, Katagiri T, Kondoh Y, Amemiya T, Tashiro H, Okazaki Y (2009) miR-210 promotes osteoblastic differentiation through inhibition of AcvR1b. FEBS Lett 583:2263–2268

Nottrott S, Simard MJ, Richter JD (2006) Human let-7a miRNA blocks protein production on actively translating polyribosomes. Nat Struct Mol Biol 13:1108–1114

Orom UA, Nielsen FC, Lund AH (2008) MicroRNA-10a binds the 5′UTR of ribosomal protein mRNAs and enhances their translation. Mol Cell 30:460–471

Petersen CP, Bordeleau ME, Pelletier J, Sharp PA (2006) Short RNAs repress translation after initiation in mammalian cells. Mol Cell 21:533–542

Pillai RS, Bhattacharyya SN, Artus CG, Zoller T, Cougot N, Basyuk E, Bertrand E, Filipowicz W (2005) Inhibition of translational initiation by Let-7 MicroRNA in human cells. Science 309:1573–1576

Pillai RS, Bhattacharyya SN, Filipowicz W (2007) Repression of protein synthesis by miRNAs: how many mechanisms? Trends Cell Biol 17:118–126

Pineles BL, Romero R, Montenegro D, Tarca AL, Han YM, Kim YM, Draghici S, Espinoza J, Kusanovic JP, Mittal P, Hassan SS, Kim CJ (2007) Distinct subsets of microRNAs are expressed differentially in the human placentas of patients with preeclampsia. Am J Obstet Gynecol 196(261):e261–e266

Pulkkinen K, Malm T, Turunen M, Koistinaho J, Yla-Herttuala S (2008) Hypoxia induces microRNA miR-210 in vitro and in vivo ephrin-A3 and neuronal pentraxin 1 are potentially regulated by miR-210. FEBS Lett 582:2397–2401

Rane S, He M, Sayed D, Vashistha H, Malhotra A, Sadoshima J, Vatner DE, Vatner SF, Abdellatif M (2009) Downregulation of miR-199a derepresses hypoxia-inducible factor-1alpha and Sirtuin 1 and recapitulates hypoxia preconditioning in cardiac myocytes. Circ Res 104:879–886

Reinhart BJ, Slack FJ, Basson M, Pasquinelli AE, Bettinger JC, Rougvie AE, Horvitz HR, Ruvkun G (2000) The 21-nucleotide let-7 RNA regulates developmental timing in *Caenorhabditis elegans*. Nature 403:901–906

Remke M, Pfister S, Kox C, Toedt G, Becker N, Benner A, Werft W, Breit S, Liu S, Engel F, Wittmann A, Zimmermann M, Stanulla M, Schrappe M, Ludwig WD, Bartram CR, Radlwimmer B, Muckenthaler MU, Lichter P, Kulozik AE (2009) High-resolution genomic profiling of childhood T-ALL reveals frequent copy-number alterations affecting the TGF-beta and PI3K-AKT pathways and deletions at 6q15-16.1 as a genomic marker for unfavorable early treatment response. Blood 114:1053–1062

Ritchie W, Flamant S, Rasko JE (2010) mimiRNA: a microRNA expression profiler and classification resource designed to identify functional correlations between microRNAs and their targets. Bioinformatics 26:223–227

Rouault TA, Tong WH (2008) Iron-sulfur cluster biogenesis and human disease. Trends Genet 24:398–407

Slack FJ, Basson M, Liu Z, Ambros V, Horvitz HR, Ruvkun G (2000) The lin-41 RBCC gene acts in the C. elegans heterochronic pathway between the let-7 regulatory RNA and the LIN-29 transcription factor. Mol Cell 5:659–669

Taguchi A, Yanagisawa K, Tanaka M, Cao K, Matsuyama Y, Goto H, Takahashi T (2008) Identification of hypoxia-inducible factor-1 alpha as a novel target for miR-17-92 microRNA cluster. Cancer Res 68:5540–5545

Thomson JM, Newman M, Parker JS, Morin-Kensicki EM, Wright T, Hammond SM (2006) Extensive post-transcriptional regulation of microRNAs and its implications for cancer. Genes Dev 20:2202–2207

Tong WH, Rouault TA (2006) Functions of mitochondrial ISCU and cytosolic ISCU in mammalian iron-sulfur cluster biogenesis and iron homeostasis. Cell Metab 3:199–210

Tsai NP, Lin YL, Wei LN (2009) MicroRNA mir-346 targets the 5′-untranslated region of receptor-interacting protein 140 (RIP140) mRNA and upregulates its protein expression. Biochem J 424:411–418

Vasudevan S, Steitz JA (2007) AU-rich-element-mediated upregulation of translation by FXR1 and Argonaute 2. Cell 128:1105–1118

Vasudevan S, Tong Y, Steitz JA (2007) Switching from repression to activation: microRNAs can upregulate translation. Science 318:1931–1934

Vasudevan S, Tong Y, Steitz JA (2008) Cell-cycle control of microRNA-mediated translation regulation. Cell Cycle 7:1545–1549

Volinia S, Calin GA, Liu CG, Ambs S, Cimmino A, Petrocca F, Visone R, Iorio M, Roldo C, Ferracin M, Prueitt RL, Yanaihara N, Lanza G, Scarpa A, Vecchione A, Negrini M, Harris CC, Croce CM (2006) A microRNA expression signature of human solid tumors defines cancer gene targets. Proc Natl Acad Sci USA 103:2257–2261

Wakiyama M, Takimoto K, Ohara O, Yokoyama S (2007) Let-7 microRNA-mediated mRNA deadenylation and translational repression in a mammalian cell-free system. Genes Dev 21:1857–1862

Wang K, Zhang S, Marzolf B, Troisch P, Brightman A, Hu Z, Hood LE, Galas DJ (2009a) Circulating microRNAs, potential biomarkers for drug-induced liver injury. Proc Natl Acad Sci USA 106(11):4402–4407

Wang Y, Keys DN, Au-Young JK, Chen C (2009b) MicroRNAs in embryonic stem cells. J Cell Physiol 218:251–255

Wightman B, Ha I, Ruvkun G (1993) Posttranscriptional regulation of the heterochronic gene lin-14 by lin-4 mediates temporal pattern formation in *C. elegans*. Cell 75:855–862

Xiao C, Srinivasan L, Calado DP, Patterson HC, Zhang B, Wang J, Henderson JM, Kutok JL, Rajewsky K (2008) Lymphoproliferative disease and autoimmunity in mice with increased miR-17-92 expression in lymphocytes. Nat Immunol 9:405–414

Yan HL, Xue G, Mei Q, Wang YZ, Ding FX, Liu MF, Lu MH, Tang Y, Yu HY, Sun SH (2009) Repression of the miR-17-92 cluster by p53 has an important function in hypoxia-induced apoptosis. EMBO J 28(18):2719–2732

Yanaihara N, Caplen N, Bowman E, Seike M, Kumamoto K, Yi M, Stephens RM, Okamoto A, Yokota J, Tanaka T, Calin GA, Liu CG, Croce CM, Harris CC (2006) Unique microRNA molecular profiles in lung cancer diagnosis and prognosis. Cancer Cell 9:189–198

Yu F, Yao H, Zhu P, Zhang X, Pan Q, Gong C, Huang Y, Hu X, Su F, Lieberman J, Song E (2007) let-7 regulates self renewal and tumorigenicity of breast cancer cells. Cell 131:1109–1123

Zhang Z, Sun H, Dai H, Walsh RM, Imakura M, Schelter J, Burchard J, Dai X, Chang AN, Diaz RL, Marszalek JR, Bartz SR, Carleton M, Cleary MA, Linsley PS, Grandori C (2009) MicroRNA miR-210 modulates cellular response to hypoxia through the MYC antagonist MNT. Cell Cycle 8:2756–2768

Zhu XM, Han T, Sargent IL, Yin GW, Yao YQ (2009) Differential expression profile of micro-RNAs in human placentas from preeclamptic pregnancies vs normal pregnancies. Am J Obstet Gynecol 200(661):e661–e667

Oxygen Sensing: A Common Crossroad in Cancer and Neurodegeneration

Annelies Quaegebeur and Peter Carmeliet

Contents

Abstract Prolyl hydroxylase domain (PHD) proteins are cellular oxygen sensors that orchestrate an adaptive response to hypoxia and oxidative stress, executed by hypoxia-inducible factors (HIFs). By increasing oxygen supply, reducing oxygen consumption, and reprogramming metabolism, the PHD/HIF pathway confers tolerance towards hypoxic and oxidative stress. This review discusses the involvement of the PHD/HIF response in two, at first sight, entirely distinct pathologies with opposite outcome, i.e. cancer leading to cellular growth and neurodegeneration resulting in cell death. However, these disorders share common mechanisms of

A. Quaegebeur and P. Carmeliet (✉)
Vesalius Research Center (VRC), VIB, K.U. Leuven, Campus Gasthuisberg, Herestraat 49, 3000 Leuven, Belgium
e-mail: peter.carmeliet@vib-kuleuven.be

M. Celeste Simon (ed.), *Diverse Effects of Hypoxia on Tumor Progression*,
Current Topics in Microbiology and Immunology 345, DOI 10.1007/82_2010_83
© Springer-Verlag Berlin Heidelberg 2010, published online: 26 June 2010

sensing oxygen and oxidative stress. We will focus on how PHD/HIF signaling is pathogenetically implicated in metabolic and vessel alterations in these diseases and how manipulation of this pathway might offer novel treatment opportunities.

Abbreviations

AMPK	Adenosine monophosphate-activated protein kinase
Ang2	Angiopoietin 2
BNIP3	Bcl-2/adenovirus E1B 19-kDa-interacting protein 3
Bv8	Bombina variagata peptide 8
CBP	CREB-binding protein
CXCR4	CXC chemokine receptor 4
FDG-PET	Fluoro deoxy glucose-positron emission tomography
FGF2	Fibroblast growth factor 2
Flt1	Fms-like tyrosine kinase 1
LDH	Lactate dehydrogenase
LRP	Low-density lipoprotein receptor-related protein
MMP9	Matrix metalloproteinase 9
mTOR	(Mammalian) target of rapamycin
NF-κB	Nuclear factor-kappaB
PDGF	Platelet-derived growth factor
PDH	Pyruvate dehydrogenase
PDK	PDH kinase
PI3	Phosphatidylinositol 3
PlGF	Placental growth factor
PTEN	Phosphatase and tensin homologue deleted on chromosome ten
PUMA	p53 up-regulated modulator of apoptosis
SDF-1	Stromal-derived factor 1
TCA	Tricyclic acid
uPAR	Urokinase-type plasminogen activator receptor
VEGFR2	Vascular endothelial growth factor receptor 2

1   Introduction

Since the appearance of oxygen in the atmosphere billions of years ago, eukaryotic life has become critically dependent on aerobic metabolism (Semenza 2007). Molecular oxygen (O_2) serves as terminal electron acceptor in a process referred to as oxidative phosphorylation, providing the cell with a highly efficient means of energy production. However, given the reactive nature of oxygen, mitochondrial oxidative metabolism exposes the cell also to the threat of reactive oxygen species (ROS). Both decreased oxygen levels (hypoxia) and oxidative stress, resulting from excessive ROS generation, posed some of the greatest evolutionary challenges for aerobic life (Taylor and Pouyssegur 2007). Therefore, oxygen-consuming organisms have developed a variety of oxygen sensing and adaptive systems, allowing a tightly regulated oxygen homeostasis. Mammalian cells are equipped with an oxygen sensor class of 2-oxoglutarate dependent iron(ii)-dioxygenases consisting of three different forms of prolyl hydroxylase domain proteins (PHD1-3) and a single asparaginyl hydroxylase, called factor inhibiting HIF (FIH) (Kaelin and Ratcliffe 2008). These oxygen sensors direct the accumulation and activation of the hypoxia-inducible transcription factors (HIF-1α, -2α and -3α), acting as master switches of the hypoxia transcriptional response (Kaelin and Ratcliffe 2008).

This endogenous protective mechanism has received great medical attention in recent years as hypoxia-related stress and deregulated oxygen homeostasis are commonly implicated in diverse disease states, with cancer and neurodegeneration being the scope of our review. These two different disease processes are characterized by a fundamentally opposing cell fate: whereas a tumor represents an uncontrolled cell proliferation escaping tissue homeostasis, neurodegenerative disorders refer to a progressive dysfunction and dying of selective neuronal subpopulations. However, these pathologies share an intriguing overlap in how their disease course is affected by hypoxia signaling. We will first describe the evidence for the role of hypoxia in the pathogenesis of both diseases, and then discuss the molecular mediators of the cellular hypoxic response as well as their biological role in both cancer and neurodegeneration. As cellular oxygen levels result from the balance between oxygen supply and consumption, we will further focus on the link between hypoxia, hypoxia signaling, blood vessels, and cellular metabolism in these diseases. We will illustrate how deciphering this link may yield new insights in the molecular etiology and offer potential therapeutic targets as well.

2   Hypoxia in Disease

2.1   Hypoxia in Cancer

Hypoxia is a well-established feature of nearly all solid tumors (Bertout et al. 2008). Tumor cells grow in multiple layers around blood vessels, resulting in a spatially highly heterogeneous distribution of oxygen levels within the tumor microenvironment (Fig. 1). Cells located at a distance of more than 100–200 μm away from blood vessels, which corresponds to the distance limit in oxygen diffusion (Gatenby and Gillies 2004; Helmlinger et al. 1997), suffer life threatening low amounts of oxygen, which may lead to necrotic tumor areas. The rapidly expanding tumor mass is craving for oxygen and therefore induces the formation of new blood vessels. Tumor angiogenesis can, however, not keep up with the increased oxygen demand and tumor cells often outgrow their vascular supply, resulting in a so-called "diffusion-limited hypoxia" (Harris 2002; Vaupel and Mayer 2007). Additionally, newly formed

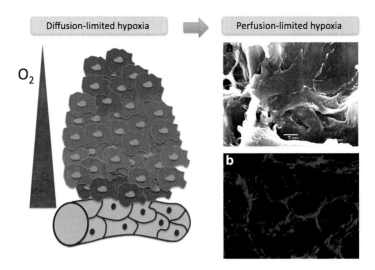

Fig. 1 Hypoxia and perfusion in a tumor. Tumor cells become hypoxic as they outgrow their blood supply. This "diffusion-limited hypoxia" drives an excessive angiogenic response, which results in a disorganized, highly abnormal vasculature. Scanning electron microscopy (**a**) and whole mount tumor sections for CD31 (**b**; in *red*) illustrate the hyperactive, pseudostratified and discontinuous tumor vasculature, further compromising vessel perfusion, referred to as "perfusion-limited hypoxia". Figure (**a** and **b**) reprinted from (Mazzone et al. 2009), with permission from Elsevier

tumor vessels are chaotically organized, immature and unstable (Fig. 1), leading to hypoperfusion and poor oxygenation; the interstitial hypertension due to the abnormal vessel leakiness induces tumor vessel collapse, thereby further enforcing hypoperfusion (Carmeliet and Jain 2000; Jain 2005). Moreover, sudden perfusion changes because of transiently occluded and reperfused vessels and abnormal rheology of hyperviscous blood give rise to an excessive generation of ROS (Dewhirst et al. 2008; Guzy and Schumacker 2006). This inefficient oxygenation is referred to as "perfusion-limited hypoxia" (Harris 2002; Vaupel and Mayer 2007).

Hypoxia is almost dogmatically expected to limit tumor growth. Nevertheless, a large amount of clinical and experimental evidence shows that poorly oxygenated tumors display a more malignant phenotype characterized by invasiveness, metastasis, hypoxia tolerance, and angiogenesis (Bertout et al. 2008; Brown and Wilson 2004) with poor treatment response and worse prognosis (Vaupel and Mayer 2007; Wouters et al. 2004). As we will outline later, exploitation of the cellular adaptive responses to hypoxia underlies this aggressive tumor behavior.

2.2 Hypoxia in Neurodegeneration

The brain is particularly sensitive to hypoxia and oxidative stress (Acker and Acker 2004). This vulnerability is mainly due to the extraordinary metabolic requirements of normally functioning neurons. Vital neuronal processes such as neurotransmission and ion homeostasis are critically depending on a continuous oxygen and glucose supply (Acker and Acker 2004). An intimate structural and functional interaction between endothelial cells, pericytes, glia and neurons – coined the blood–brain barrier (BBB) in the neurovascular unit – serves a neurovascular crosstalk, allowing the nervous system to match blood supply to the changing neuronal energy needs (Iadecola 2004; Iadecola and Nedergaard 2007; Zlokovic 2008). The homeostasis of the cerebral microenvironment is maintained through a tightly sealed endothelial layer that limits free transport of molecules across the BBB. Any cerebrovascular deficit or neurovascular uncoupling, even subtle in nature, may therefore lead to metabolic deregulation, ultimately progressing to metabolic collapse and neuronal death (Iadecola 2004). Oxygen radicals are highly reactive species that are generated as normal byproducts of mitochondrial oxidative metabolism (Kulkarni et al. 2007). In physiological conditions, antioxidant enzymes keep their levels in check. Hypoxia may cause oxidative stress, as mitochondria generate more ROS in hypoxic conditions (Guzy and Schumacker 2006). Given the limited antioxidant defense mechanisms, especially in the brain, excessive ROS formation can result in oxidative damage due to peroxidation of proteins, lipids, RNA, and DNA (Andersen 2004; Kulkarni et al. 2007).

Hence, both perfusion deficits and oxidative stress are important mediators of neuronal dysfunction and death. Not surprisingly, vascular abnormalities, neurovascular uncoupling and free radical injury have been observed in the neurodegenerative brain (Andersen 2004; Segura et al. 2009; Storkebaum and Carmeliet 2004; Zlokovic 2008). Where these processes should be situated in the chain of pathologic events, i.e. whether they are cause or consequence, remains an outstanding debated question. However, the evidence for vessels and metabolism as active contributors to progressive neuronal degeneration is ever rising, setting the scene for exploring the role of hypoxia signaling in neurodegenerative disorders.

3 Hypoxia Signaling Pathways

3.1 The Molecular Players of Hypoxia Signaling

Molecular oxygen represents a vital source of energy for most eukaryotic organisms. During embryonic development as well as in various (patho)-physiological conditions, cells are often exposed to varying oxygen levels. Mammalian organisms are thus equipped with intricate oxygen sensing mechanisms instructing a variety of adaptive responses, in an attempt to maintain their energetic balance. In this review, we will focus on the prolyl hydroxylase domain proteins (PHD1-3) and factor inhibiting HIF (FIH) (Aragones et al. 2009; Kaelin and Ratcliffe 2008). These cellular oxygen sensors activate in an oxygen-dependent manner a major transcriptional pathway governed by the hypoxia-inducible factors, allowing the cell to transduce oxygen levels to adaptive gene expression. The prolyl hydroxylase domain proteins (PHDs) hydroxylate specific proline residues of HIFα, which are recognized by the Von Hippel Lindau (VHL) protein. The latter recruits an ubiquitin ligase complex, targeting HIFα for proteosomal degradation (Kaelin and Ratcliffe 2008; Schofield and Ratcliffe 2004). FIH carries out the hydroxylation of a specific asparagine residue, interfering with the recruitment of essential transcriptional co-activators such as p300 and CBP (Kaelin and Ratcliffe 2008; Schofield and Ratcliffe 2004). PHDs and FIH act as *bona fide* oxygen sensors because they use molecular oxygen, besides 2-oxoglutarate, as substrate in the hydroxylation reaction. Consequently, when oxygen levels drop, these oxygen sensors start to lose their activity. This results in accumulation and transcriptional activation of HIFα in a complex with its constitutively expressed counterpart HIFβ, binding at the hypoxia-response element (HRE) in the promoter of numerous genes (Semenza 2003).

To provide the necessary fine-tuning of the hypoxia response and shape a cellular outcome suited for the varying contexts of severity and duration of the hypoxic stimulus, the PHD/HIF pathway exhibits a high degree of flexibility and complexity (Lendahl et al. 2009). This diversity is in part achieved by the differing oxygen affinities of PHDs and FIH, and the specific transcriptional programs induced by each HIF isoform (Kaelin and Ratcliffe 2008). In addition, HIF-independent targets of PHDs are being increasingly identified, among which members of the canonical NF-κB activation pathway (Chan et al. 2009; Cummins et al. 2006; Xue et al. 2010), Notch (Coleman et al. 2007; Zheng et al. 2008), RNA polymerase II (Mikhaylova et al. 2008) and others. Moreover, there is emerging evidence for an intimate crosstalk of the PHD/FIH/HIF pathway with other major gene regulatory pathways, intersecting with diverse biological processes, such as with nutrient metabolism via regulation of AMPK, with protein synthesis, folding and degradation via effects on mTOR and the unfolded protein response (Wouters and Koritzinsky 2008), with apoptosis and cell survival via p53 (Xenaki et al. 2008), and many others. Finally, another level of complexity is added by the posttranslational modifications of HIF, as well as transcriptional regulation by microRNA and epigenetic factors (Lendahl et al. 2009).

In vertebrates, an elaborate vessel network conducts oxygen and glucose to every cell, where these energy substrates enter the metabolic process of ATP generation. Obviously, any disturbance in vascular supply or cellular metabolism will threat the energy homeostasis and trigger adaptive processes adjusting oxygen supply to its demand and vice versa (Aragones et al. 2009). PHD signaling as described above enables hypoxic cells to survive and function by initiating programs that enhance oxygen supply via "cell extrinsic" mechanisms on one hand, and that reduce energy expenditure and oxygen consumption via "cell intrinsic" mechanisms on the other hand (Aragones et al. 2009) (Fig. 2).

3.2 Hypoxia Signaling in Cancer

Both HIF-1α and HIF-2α have been implicated in the hypoxia response of tumors, providing cancer cells a survival advantage in conditions of varying oxygen levels

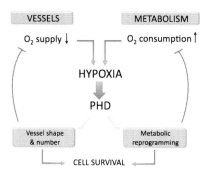

Fig. 2 Vascular and metabolic in- and output of the PHD cascade. Many physiological and pathological conditions are associated with an imbalance between oxygen supply and consumption. The PHD molecules will sense the resulting hypoxia and induce an adaptive feedback at the level of vessels and cellular metabolism by increasing oxygen supply and reducing oxygen consumption respectively, in order to restore oxygen homeostasis

(Brown and Wilson 2004; Harris 2002; Pouyssegur et al. 2006; Rankin and Giaccia 2008; Semenza 2003). Consistent with this note, HIF-1α and HIF-2α levels are upregulated in several tumor specimens and associate with a poor prognosis and therapy resistance (Rankin and Giaccia 2008; Semenza 2010a; Sullivan and Graham 2009). Although the majority of research supports a tumor promoting function of both HIF isoforms, accumulating evidence also points to a tumor-suppressive role depending on the tumor context and cell type (Rankin and Giaccia 2008). For instance, HIF-1α deficiency in embryonic stem cells protects cells from hypoxia-induced apoptosis and accelerates tumor growth (Carmeliet et al. 1998). Similarly, HIF-2α-overexpressing glioblastoma tumors display reduced tumor growth and increased apoptosis (Acker et al. 2005).

The role of the cellular oxygen sensors in tumor physiology has been less extensively studied so far and remains incompletely understood. As negative regulators of HIFs, PHDs are expected to act as tumor suppressors. Accordingly, PHD activity is reduced in several tumor cell lines (Calvisi et al. 2007; Chan et al. 2002; Kato et al. 2006; Knowles et al. 2003). Increased PHD expression levels are reported by some (Couvelard et al. 2008; Jokilehto et al. 2006), but not by all groups (Yan et al. 2009b). In some studies, elevated PHD2 levels correlate with tumor aggressiveness (Couvelard et al. 2008; Jokilehto et al. 2006) and radiation resistance (Luukkaa et al. 2009). Others document that PHD2 inhibition causes chemoresistance due to HIF-1α induced expression of the multi-drug resistance gene (Brokers et al. 2010) and promotes tumor growth through HIF-independent effects on NF-κB (Chan et al. 2009), while PHD activation impairs tumor growth (Matsumoto et al. 2009; Qi et al. 2008; Tennant et al. 2009). As a HIF-target, PHD2 is involved in a negative feedback loop to avoid overactivation of HIFs in the hypoxic tumor microenvironment (Henze et al. 2010; Marxsen et al. 2004; Metzen et al. 2003), providing an explanation for the discrepancy between PHD activity and expression levels. Furthermore, a biphasic role of PHD2, with a tumorigenic effect depending on its level, was recently suggested, underscoring the complex consequences of PHD activity in tumor cells (Lee et al. 2008). The role of the other PHDs has been even less well studied, but may be non-overlapping. For instance, PHD1 inhibition impedes tumor breast cell proliferation through cyclin D1 expression (Zhang et al. 2009). PHD3 seems to counteract the tumor-suppressive activity of HIFs in glioma models (Henze et al. 2010). In colorectal cancer, by contrast, downregulated PHD3 levels associate with malignant behavior (Xue et al. 2010). These results suggest that the overall cancer activity of PHDs is contextual and dependent on the type, stage, and treatment response of tumors.

Genetic alterations in the hypoxia pathway also drive expression of HIF target genes. For instance, in Von Hippel Lindau (VHL) disease, a hereditary cancer syndrome caused by germline mutations in the VHL gene, normoxic degradation of HIFα is blocked. The accumulating HIF levels result in the development of highly vascularized tumors (Kaelin 2008). Loss of function mutations in PHD2 also cause tumorigenesis (Kato et al. 2006). Additionally, activation of oncogenes and loss of tumor suppressors functioning in growth factor signaling pathways contribute to HIF-driven gene expression as well, mainly through increased HIF-1 synthesis (Semenza 2003, 2010a). Taken together, the hypoxia response is a crucial mediator of tumor malignancy.

3.3 Hypoxia Signaling in Neurodegeneration: Lessons from Hypoxic Preconditioning and Ischemic Tolerance

Hypoxia and oxidative stress can, when prolonged and severe, result in neuronal death. However, it has been known for some time that exposing the brain to a controlled stressful stimulus, such as oxygen deprivation, to an extent that neuronal function is slightly impaired, yet not irreversibly damaged, elicits a protective response. In this way, the brain acquires a state of ischemic or hypoxia tolerance and will be protected against a subsequent lethal stimulus (Gidday 2006). This phenomenon is known as ischemic brain preconditioning and has been documented in other organs as well. This concept is of great medical interest as it might prove a preventive strategy in high-risk conditions for cerebrovascular disease such as a transient ischemic attack or subarachnoid hemorrhage (Dirnagl et al. 2009). At a first glance, hypoxic preconditioning is not implicated in the field of neurodegenerative disorders given the chronic nature of the latter. Nevertheless, an in-depth molecular understanding of preconditioning may allow the identification of novel neuroprotective players, which could represent attractive disease candidates in neurodegeneration. Moreover, stroke and neurodegenerative diseases share common pathophysiological processes preceding neuronal death.

A preconditioning, non-lethal, stimulus is known to profoundly alter transcriptional output in the brain (Bernaudin et al. 2002; Stenzel-Poore et al. 2003). The HIF family represents one of the most studied hypoxia-sensitive transcription factors involved in this genetic reprogramming (Correia and Moreira 2010; Harten et al. 2010). Both *in vitro* and *in vivo* studies show increased levels of HIF-1α and its downstream targets VEGF, erythropoietin and glycolytic enzymes upon exposure to cellular stress (Baranova et al. 2007; Sharp and Bernaudin 2004). Given the well-documented neurotrophic effects of the effector molecules VEGF (Ruiz de Almodovar et al. 2009) and erythropoietin (Brines and Cerami 2005), HIF-1α is arguably pinpointed as moderator of neuroprotection. Further *in vivo* characterization of the role of HIF-1α in acute cerebral ischemia yields, however, conflicting data (Baranova et al. 2007; Chen et al. 2009a; Helton et al. 2005). Pharmacological manipulation of HIF-1α also demonstrates diverging effects on stroke outcome (Baranova et al. 2007; Chen et al. 2008, 2009a; Ratan et al. 2008; Siddiq et al. 2005; Zhou et al. 2008). This may be related to the dual activity of HIF-1α, which not only activates survival-promoting pathways, but also triggers pro-apoptotic proteins such as BNIP3 and PUMA (Chen et al. 2009b). Similar to the diverging role of HIF-1α in cancer, the overall effect of HIF-1α on neuronal cell fate appears to be context specific.

As HIF-1α levels are importantly regulated by the activity of the PHDs, inhibition of PHDs has become an attractive strategy in preconditioning therapy. Current available PHD inhibitors act as iron chelators or 2-oxoglutarate analogs, depleting the enzyme cofactor iron or competing with its cosubstrate 2-oxoglutarate, respectively (Fraisl et al. 2009). The neuroprotective effects of these agents have been validated in preclinical stroke models (Baranova et al. 2007; Freret et al. 2006; Li et al. 2008; Sharp and Bernaudin 2004; Siddiq et al. 2005), with substantial evidence for a, at least partly, HIF-1α-mediated effect (Baranova et al. 2007; Hamrick et al. 2005).

Inspired by the neuroprotective role of the PHD/HIF pathway in brain preconditioning and ischemic tolerance, PHD inhibitors are assessed as treatment design for neurodegenerative disorders as well. Already in 1991, clinical efficacy of iron chelators was tested in AD patients (Crapper McLachlan et al. 1991). Desferrioxamine, another iron chelator, was reported to slow down neurodegeneration in rodent models of Parkinson's disease (Ben-Shachar et al. 1991; Lan and Jiang 1997). Only recently, novel iron chelating drugs were assessed in ALS and resulted in a survival advantage in mutant SOD1 mice (Kupershmidt et al. 2009). The lack of specificity limits the therapeutic utility of the current PHD inhibitors and confounds the role of specific PHD isoforms in neurodegeneration. While gene deletion and transient knockdown studies of PHD1 in skeletal muscle and liver, and of PHD2 in heart show protection against ischemic injury (Aragones et al. 2008; Eckle et al. 2008; Hyvarinen et al. 2010; Schneider et al. 2009), the isoform-specific role of PHDs in different neural cell types in the context of neurodegeneration remains to be largely explored. So far, a single *in vitro* study implicated PHD1 in the neuroprotective response to oxidative stress (Siddiq et al. 2009). Obviously, complementing *in vivo* analyses are warranted to unravel the disease- and cell-specific role of PHDs in neurodegenerative disorders.

Taken together, hypoxia elicits through activation of the PHD/HIF pathway a transcriptional response, which both in the brain as in a tumor generally confers protection against hypoxic stress, though context-dependent divergent effects have been documented as well. The broad variety of biological processes involved in this adaptive response ranges from angiogenesis, metabolism, apoptosis, cell stem physiology, inflammation, autophagy, endoplasmatic reticulum stress, proteosomal homeostasis, etc. Interestingly, each of these processes has been implicated in the pathogenesis of cancer and neurodegenerative disorders. Thus, an extensive intersection with the hypoxia signaling machinery emerges at the level of several pathobiological events, representing an intriguing overlap of these divergent disease states. With oxygen delivery and consumption as ultimate determinants of the hypoxia response, we will limit our further discussion to the contribution of vessel and metabolic alterations to the pathogenesis of cancer and neurodegeneration, in particular highlighting the involvement of hypoxia and the PHD/HIF pathway.

4 Angiogenesis

4.1 Angiogenesis in Cancer: Targeting Vessel Numbers or Endothelial Shape?

The ability to instruct the formation of new blood vessels is a prerequisite for cancer to manifest as an aggressively expanding and infiltrating cell mass (Hanahan and Weinberg 2000). Hypoxia strongly triggers the "angiogenic switch" during tumorigenesis by inducing the expression of numerous angiogenic cytokines, in an attempt to meet the increasing oxygen and nutritional need of the highly metabolic tumor (Carmeliet and Jain 2000; Hanahan and Folkman 1996). Furthermore, direct or indirect activation of HIFs by oncogenes or pro-tumorigenic signaling pathways additionally drives the "angiogenic switch" (Carmeliet and Jain 2000). Importantly, however, and somewhat paradoxically, such excessive release of pro-angiogenic factors in cancer generates highly abnormal, disorganized, immature tumor vessels (Greenberg et al. 2008; Jain 2005), that impair perfusion, thereby further aggravating tumor hypoxia and fueling a self-reinforcing vicious cycle of a non-productive angiogenic response (Fig. 3).

Based on the presumption that tumors require blood supply to grow, efforts have been undertaken to block this supply as maximally as possible, with the hope to eradicate hereby the tumor by starving it to death (Folkman 1971). Today, anti-angiogenic drugs of the VEGF signaling pathway are most widely used in clinical oncology practice (Ellis and Hicklin 2008; Folkman 2007). Despite successes, the tumor suppressive effect of anti-angiogenic therapy often appears short-lived, translating in an overall modest effect on progression-free and overall survival for several cancers (Bergers and Hanahan 2008). This transitory efficacy suggests that anti-VEGF therapy inflicts adaptive responses in the tumor and surrounding stromal cells. Various molecular and cellular mechanisms have been implicated as possible causes to explain this evasive treatment resistance (Bergers and Hanahan 2008;

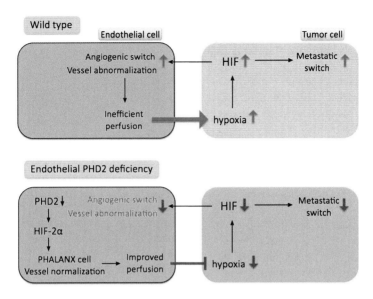

Fig. 3 Endothelial PHD2 deficiency mediates endothelial normalization in tumor vessels. (**a**) Hypoxia in a tumor induces the HIF-driven expression of genes that promote angiogenesis and metastasis. The excessive release of angiogenic cytokines results in tumor vessel abnormalization and dysfunctional perfusion, reinforcing tumor hypoxia. (**b**) In mice lacking one allele of PHD2 in endothelial cells, the endothelial phenotype is directed towards the "phalanx cell", resulting in a tightly aligned, quiescent endothelial layer. The improved perfusion that goes along with endothelial normalization will alleviate tumor hypoxia, attenuating HIF-driven expression of angiogenic and prometastatic genes

Carmeliet 2005; Loges et al. 2009). We will here discuss how the hypoxia response may contribute to this process.

Blocking VEGF signaling inhibits neoangiogenesis and prunes existing vessels, leading to a decline of tumor oxygen levels (Franco et al. 2006). By inducing hypoxia, anti-angiogenic agents can thus select hypoxia resistant and intrinsically more malignant tumor clones (Graeber et al. 1996; Yu et al. 2002). Additionally, hypoxia-driven expression of other pro-angiogenic factors such as Ang2, PDFG-A, FGF2, and PlGF upon anti-VEGF treatment results in tumor revascularization (Casanovas et al. 2005; Fernando et al. 2008). The recruitment of pro-angiogenic myeloid cells to the hypoxic tumor environment, in part resulting from HIF-upregulated SDF-1 levels, further contributes to this mechanism of resistance (Du et al. 2008; Shaked et al. 2006). Thus, hypoxia signaling has a central role in driving the secondary wave of angiogenesis that is iatrogenically induced and may limit the success of the initial anti-angiogenic treatment strategy. Alternative angiogenesis signaling circuitries, many of which are induced by hypoxia, are therefore currently receiving attention as they represent attractive novel targets for anti-angiogenic therapy: amongst those are antibodies targeted against Bv8, neuropilin-1, PlGF, PDGF-C, and others (Crawford et al. 2009; Fischer et al. 2008;

Folkman 2007). Whether combining anti-angiogenic strategies may overcome treatment resistance remains clinically largely unanswered.

Besides stimulating tumor revascularization, hypoxia in response to anti-angiogenic therapy may also incite two other processes, i.e. tumor cell invasiveness and metastasis, that may ultimately cause much greater challenges to resolve therapeutically (Bergers and Hanahan 2008). Two recent studies documented increased invasiveness and metastatic disease upon pharmacological or genetic abrogation of VEGF signaling in distinct mouse models (Ebos et al. 2009; Paez-Ribes et al. 2009). Anti-VEGF treatment has also been found to induce pro-metastatic genes such as SDF-1 and CXCR4 (Xu et al. 2009). The question to what extent these preclinical findings are either due to experimental conditions or are clinically relevant is not easily addressed, considering other evidence for a suppressive effect of VEGF-inhibitors on metastasis (Crawford and Ferrara 2009). At least in glioblastoma multiforme, multifocal recurrence was observed in some studies upon anti-VEGF therapy (Narayana et al. 2009; Norden et al. 2008; Zuniga et al. 2009). Regardless of these outstanding questions, it is worthwhile noting that, even though tumor cells have acquired means to survive severe hypoxic stress conditions, oxygen levels may be extremely low inside the tumor stroma, approaching anoxia. This hostile oxygen-deprived microenvironment forces tumor cells to escape, resulting in increased invasiveness and metastasis. Emerging evidence indicates that hypoxia signaling may represent an important molecular signature of this invasive/metastatic tumor phenotype. Indeed, hypoxia fuels infiltrative and metastatic tumor cell behavior at different biological levels. Stimulation of the epithelial to mesenchymal transition, expression of the hepatocyte growth factor receptor c-Met, MMP9, uPAR are only a few molecular examples that constitute the dynamic link between hypoxia, HIF-mediated expression and invasiveness (Loges et al. 2009; Pouyssegur et al. 2006; Sullivan and Graham 2007). To which extent the hypoxia response actually contributes to the exacerbated invasive and metastatic behavior upon VEGF inhibition in cancer remains to be further explored.

Recent genetic studies yielded insights that targeting a key oxygen-sensing mechanism whereby endothelial cells regulate their ancient function to secure oxygen supply and organ perfusion may provide alternative anti-vascular strategies to combat cancer (Mazzone et al. 2009) (Fig. 3). This strategy relies on targeting the endothelial shape rather than the numbers of tumor vessels. Indeed, in tumor-bearing mice lacking a single PHD2 allele in endothelial cells, the endothelial layer in tumor vessels appeared more regular, tight and quiescent, and was surrounded by pericytes and a basement membrane, all together improving vessel stability and maturation, and leading to increased blood conduction. Importantly, invasive growth and metastasis were significantly impeded. This less malignant phenotype results in part from an improved barrier function of the endothelial layer, preventing the escape of tumor cells towards the vascular lumen, together with an alleviated hypoxia-driven genetic program. Notably, the microvessel density was unchanged, calling into question the relevance of using microvascular density as sole biomarker of tumor angiogenesis. At the cellular level, tumor vessel normalization was mediated by a change of endothelial cells to a phalanx phenotype, a quiescent non-proliferating and non-migrating endothelial cell aligned in a tight cobblestone layer (Mazzone et al. 2009). Upregulated levels of soluble Flt1 (a VEGF trap) and the junctional molecule VE-cadherin, known to decrease endothelial permeability and VEGF-mediated responses, counteract the abnormalizing effects of VEGF on the tumor vasculature (Mazzone et al. 2009).

4.2 Angiogenesis in Neurodegeneration: Hypoxia and Neurovascular Dysfunction

In many neurodegenerative diseases, vessel defects and anomalies have been documented (Storkebaum and Carmeliet 2004). Whether these vascular alterations play an active role in the pathogenesis as initiators or as modifiers has, however, not been extensively addressed. We will envision to what extent hypoxia, deregulated hypoxia signaling and vascular defects act as pathological insults to the neurovascular unit in two common neurodegenerative disorders, i.e. amyotrophic lateral sclerosis and Alzheimer's disease, and will speculate briefly about the therapeutic potential of targeting the PHD pathway for treatment of cerebrovascular dysregulation in neurodegeneration.

4.2.1 Amyotrophic Lateral Sclerosis

Amyotrophic lateral sclerosis (ALS) is characterized by a fairly selective degeneration of motor neurons in the motor cortex, brainstem and spinal cord and results in death on average within 3–5 years due to respiratory failure (Rowland and Shneider 2001). Unfortunately, current clinical practice offers no satisfying impact on disease course, which is mainly due to our poor understanding of the biology of motor neuron death. Chronic vascular insufficiency was implicated rather unexpectedly in this disease. Genetically engineered mice with a deletion of the hypoxia-response element in the VEGF promoter, resulting in a 25–40% decrease in VEGF levels, the so-called VEGF$^{\partial/\partial}$ mice, exhibited a slowly progressive motor neuron disease, reminiscent of ALS in humans (Oosthuyse et al. 2001) (Fig. 4). Besides neural hypoperfusion, an insufficient VEGF-dependent neuroprotection was postulated to contribute. Neuronal overexpressing of VEGFR2 (Storkebaum et al. 2005) and VEGF (Wang et al. 2007) has indeed been shown to improve the motor phenotype of mutant SOD1 mice, a preclinical model of ALS. Because of its pleiotropic multitasking effects in the nervous system, VEGF is designated as one of the prototypic angioneurins (Zacchigna et al. 2008). Human genetic studies in several populations confirmed an involvement of low VEGF levels in the development of ALS. In a recent meta-analysis, homozygous carriers of a particular risk VEGF allele, associated with reduced VEGF expression levels, display an increased susceptibility to ALS (Lambrechts et al. 2009). Additional support for

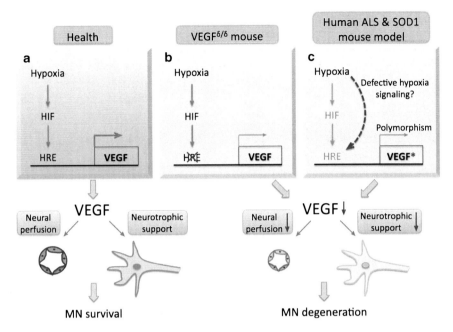

Fig. 4 Blunted hypoxia response in motor neuron disease. (**a**) In healthy conditions, the CNS will upon limited oxygen availability accumulate HIF, which will bind to a HRE in the promoter of the VEGF gene. VEGF maintains neural perfusion and provides neurotrophic support to the motor neurons, stimulating their survival. (**b**) In the VEGF$^{\delta/\delta}$ mouse, the hypoxic upregulation of VEGF in the CNS is blunted due to a genetic deletion in the HRE of the VEGF gene. Insufficient neural perfusion and neurotrophic support are likely contributive to the resulting motor neuron degeneration. (**c**) Both in human ALS and in the SOD1 mouse model, emerging evidence points to a deregulated hypoxia response as VEGF levels are reduced in hypoxic conditions. (*CNS* central nervous system, *HRE* hypoxia-response element, *SOD1* superoxide dismutase 1, *MN* motor neuron)

the relevance of VEGF in human disease comes from studies showing decreased VEGF levels in the cerebrospinal fluids of ALS patients (Devos et al. 2004).

A possible involvement of PHD/HIF signaling in deficient VEGF expression and ALS has not been addressed extensively so far. Yet, (pre)clinical evidence suggests a deregulated hypoxia response (Fig. 4). In the mutant SOD1 mouse model, there is insufficient upregulation of VEGF levels in hypoxic conditions (Murakami et al. 2003). Similar observations were made in hypoxic ALS patients (Just et al. 2007; Moreau et al. 2006, 2009). Also, angiogenin, another hypoxia-inducible angioneurin, has been genetically implicated in familial and sporadic ALS cases (Greenway et al. 2006) and was found to protect neurons against hypoxic injury (Sebastia et al. 2009).

Evaluation of the therapeutic potential of VEGF administration revealed that both intramuscular viral VEGF gene transfer with subsequent retrograde axonal transport of the viral vector as well as intracerebroventricular VEGF protein delivery improve the motor phenotype of SOD1 mice and rats (Azzouz et al. 2004; Storkebaum et al. 2005). A clinical trial assessing the safety of intracerebroventricular VEGF administration is currently ongoing. So far, these studies have resulted in the recognition of VEGF as an orphan drug for ALS patients.

Interestingly, another type of vascular defects, i.e. blood spinal cord barrier defects were demonstrated in a mutant SOD1 mouse model, before apparent signs of motor neuron degeneration (Garbuzova-Davis et al. 2007; Zhong et al. 2008). Knockdown of mutant SOD1 in endothelial cells reduced pathological BBB permeability, although there was no significant improvement of the phenotype (Zhong et al. 2009).

Thus, whereas excessive secretion of VEGF characterizes many cancers, insufficient VEGF signaling causes and contributes to neurodegeneration. An outstanding question in this regard is whether VEGF-inhibitors used as cancer therapeutics might promote or aggravate neurodegeneration. Indeed, a recent case report described oculomotor nerve palsy after intravitreal administration of VEGF targeting therapy (Micieli et al. 2009). Also, stroke-like lesions with diffusion restriction, necrosis and HIF-1α upregulation were induced by an anti-VEGF antibody within the previously enhancing tumor area in glioma patients receiving anti-angiogenic therapy (Rieger et al. 2009). Neurotoxic effects have been documented in the retina in some but not in all reports (Nishijima et al. 2007; Saint-Geniez et al. 2008).

4.2.2 Alzheimer's Disease

Alzheimer's disease (AD) is the most prevalent neurodegenerative disorder, affecting more than 30 million people worldwide and is characterized by a slowly but relentlessly progressive cognitive decline. Although synaptic dysfunction and neuronal loss is generally accepted to underlie memory loss, cerebrovascular lesions have been frequently documented in AD (Smith and Greenberg 2009; Zlokovic 2005). In fact, early last century, Alois Alzheimer himself observed vessel abnormalities in the brains of AD patients. We will dissect this complex neurovascular interplay in two directions, i.e. how Aβ induces vascular anomalies and, conversely, how cerebrovascular dysfunction can contribute to neurodegeneration.

The amyloid β-peptide (Aβ) is considered as central actor in the pathogenesis of AD. Aβ is generated upon sequential cleavage of amyloid precursor protein (APP) by the β-site of APP cleaving enzyme (BACE) and the γ secretase complex (Haass and Selkoe 2007). In the normal brain, accumulation of Aβ is prevented by clearance towards the vascular compartment (Tanzi et al. 2004). An imbalance between production and degradation, as is believed to occur in AD patients, increases Aβ levels resulting in aggregate formation in the interneuronal parenchyma. Interestingly, Aβ is also deposited in the wall of small brain arteries, a condition known as cerebral amyloid angiopathy (CAA) (Greenberg et al. 2004). The latter correlates to a broad variety of clinical phenotypes, ranging from microinfarction, cerebral hemorrhages due to vascular rupture and notably, cognitive deterioration (Greenberg et al. 2004). Besides the possible exacerbation of neuronal dysfunction in AD through microinfarction and hemorrhages (Smith and Greenberg 2009), vascular amyloid interferes with the dynamic adaptation of cerebral blood flow to regional metabolic needs (Iadecola et al. 1999; Niwa et al. 2000). Aβ has been known for some time to act as a vasoconstrictive agent (Thomas et al. 1996). Endothelium-mediated production of ROS is thought to account for this disturbed vascular reactivity (Park et al. 2005). Clinical imaging studies in early symptomatic AD patients reveal a mismatch in blood flow with brain metabolism (Bookheimer et al. 2000), further substantiating the evidence for cerebrovascular dysregulation and neurovascular uncoupling early in disease.

Vascular amyloid not only disturbs vessel reactivity but also compromises the structural integrity of the vasculature. Consistent with this note, microvessel density is reduced in AD brains (Smith and Greenberg 2009). Functional imaging in AD patients additionally confirms a cerebral hypoperfusion early in disease (de la Torre 2004). One would expect that the impaired oxygenation elicits adaptive angiogenic responses. Indeed, some groups reported upregulation of VEGF in AD (Chiappelli et al. 2006; Tarkowski et al. 2002). Nonetheless, this protective growth factor signaling might be counteracted or overwhelmed by Aβ, which has reportedly anti-angiogenic activity (Paris et al. 2004a, b). Furthermore, Aβ interacts with VEGFR2, which is believed to account for the VEGF blocking effect and impairment of the protective VEGF signaling (Patel et al. 2010). Further study is required to define the role of VEGF and insufficient angiogenic signaling in the AD brain. Taken together, Aβ affects vascular function, integrity and remodeling.

Hypoxia and cerebrovascular dysfunction may also trigger AD pathology. Epidemiologically, there is a well-established association between hypoxic conditions such as stroke and cardiovascular risk factors and the development of AD (de la Torre 2004). In line with this observation, atherosclerosis is more pronounced in large vessels from AD patients (de la Torre 2004). At the molecular level, we are only starting to understand how hypoxia influences Aβ production and clearance. Exposing rodent brain to transient or prolonged ischemia increases Aβ levels (Iadecola 2004; Sun et al. 2006; Zhang et al. 2007). Hypoxia enhances the expression of BACE (Guglielmotto et al. 2009; Sun et al. 2006; Zhang et al. 2007) and APH-1A, a subunit of the γ secretase complex (Li et al. 2009; Wang et al. 2006). Interestingly, a hypoxia-response element is present in the promoters of BACE and APH-1A (Sun et al. 2006; Wang et al. 2006), suggestive of a HIF-driven reinforcement of Aβ generation. Additionally, Aβ clearance is impaired in hypoxic conditions. Pericytes in AD brain arteries express high levels of pericytic transcription factors, i.e. serum response factor (SRF) and myocardin (MYOCD), leading to hyper contractility. Hypercontractile pericytes interfere with the dynamic adaptation of blood flow, thereby contributing to vessel hypoperfusion and impairing perivascular clearance of Aβ (Chow et al. 2007). Moreover, increased levels of SRF and MYOCD in AD brain result in downregulation of LRP (Bell et al. 2009), a key receptor of transendothelial Aβ transport. As hypoxia can induce the expression of these transcription factors, it arguably can trigger this cascade (Bell and Zlokovic 2009). In Fig. 5, a neurovascular model for AD is depicted, with Aβ injury as pivotal key event.

Angiogenesis and improved vessel function are considered to represent protective mechanisms in ischemic tolerance (Dirnagl et al. 2009; Gidday 2006).

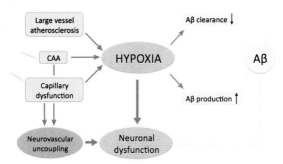

Fig. 5 Hypoxia as a central player in a feed-forward loop between Aβ metabolism and cerebro-vascular pathology in Alzheimer's disease. Besides its direct detrimental effect on neuronal function, hypoxia promotes Aβ production and accumulation. An increased deposition of Aβ in the brain parenchyma and vessel wall will affect both neuronal and vascular function. Cerebro-vascular deficits will further compromise oxygen delivery and contribute to a neurovascular uncoupling. This intersection of hypoxia with Aβ metabolism helps to explain how vessel defects can both initiate and modify AD pathology through a feed-forward loop sustained by reduced oxygen supply. (*CAA* cerebral amyloid angiopathy)

However, whether targeting PHDs might have a beneficial impact on the course of neurodegenerative disorders through vascular effects remains speculative until today. Nonetheless, certain findings may suggest to consider the vascular effects of PHD inhibition as interesting target for further study. Indeed, postnatal disruption of PHD2 augments brain angiogenesis in mice (Takeda et al. 2007) and pharmaco-logical manipulation with available PHD inhibitors upregulates VEGF levels in the brain; though the latter were insufficient to induce new vessel growth, they could still promote vessel perfusion (Siddiq et al. 2005). Whether reduced endothelial PHD2 activity leads to endothelial normalization of the pathologic vasculature in the neurodegenerative brain represents a highly puzzling question for future research.

5 Metabolism

5.1 Metabolism in Cancer: Sweet and Sour

Healthy cells with an aerobic metabolism generate ATP via oxidative phosphory-lation and, hence, when oxygen levels drop, they switch to anaerobic glycolytic metabolism to maintain energy production (Taylor and Pouyssegur 2007). Hypoxia in tumors fuels a similar metabolic program that serves tumor growth advantages.

For instance, glycolysis is stimulated through expression of glucose transporters (GLUT) and key glycolytic enzymes (Semenza 2010b), facilitated by HIF-1α upregulation. Accumulating pyruvate is shuttled away to lactate via lactate dehydrogenase A, another HIF target gene (Semenza 2010b). A cell-threatening drop in pH level is counteracted by HIF-driven expression of lactate and proton efflux transporters (Semenza 2010b). Most cancer cells also reprogram (reduce) mitochondrial respiration, oxygen consumption and consequently ROS generation. HIF-1α reduces the entry of pyruvate into the TCA cycle by inducing the expression of PDKs, which inhibit the PDH complex (Semenza 2010b). Additionally, mitochondrial autophagy is stimulated through HIF-mediated expression of BNIP3 (Semenza 2010b). HIF-1α facilitates the use of cytochrome c oxidase subunit 1 (COX4-1) over that of the COX4-2 isoform to optimize electron transport and minimize the formation of free radicals (Semenza 2010b). Cancer cells undergo other types of metabolic reprogramming, geared to promote anabolic synthesis of macromolecules that function as building blocks for rapid cell growth (Deberardinis et al. 2008; Jones and Thompson 2009; Vander Heiden et al. 2009), but it remains largely unknown whether these are under the control of hypoxia signaling and will be therefore not discussed in this review.

A similar, though not identical, metabolic reprogramming is observed in mice lacking the oxygen sensor PHD1, which provides skeletal muscle fibers hypoxia tolerance, even in the absence of any other adaptive processes governing oxygen supply such as angiogenesis, vasodilatation and erythropoiesis. Ischemic muscle necrosis upon ligation of blood supply is largely prevented by homozygous PHD1 deficiency (Aragones et al. 2008). PHD1 deficient myofibers consume less oxygen due to reduced glucose oxidation as a result of increased expression of PDKs. Increased levels of LDH and the lactate transporting monocarboxylate transporter are observed as well, facilitating glycolytic flux (unpublished results, P.C.). The hypoxia tolerant phenotype of PHD1 deficient mice is partly abrogated by HIF-2α depletion, suggesting that PHD1 acts via HIF-2α in muscle (Aragones et al. 2008). Evidence exists that PHDs control a similar metabolic reprogramming in tumor cells as well (Lee et al. 2008). Moreover, a partial "normalization" of the cellular metabolism was observed in the better-oxygenated tumors of the PHD2 deficient mice, underscoring the role of hypoxia in driving the metabolic switch (Mazzone et al. 2009).

PHDs can orchestrate tumor metabolism via additional mechanisms, other than hypoxia, such as via regulating oxidative stress. The core catalytic activity of PHDs depends on the redox state of iron, allowing the cell to respond to oxidative stress (Gerald et al. 2004). Indeed, when ROS promote oxidation of Fe^{2+}, a necessary cofactor for these oxygen-sensing enzymes, PHDs become less active and, hence, induce HIF-1α (Guzy and Schumacker 2006), which suppresses oxidative metabolism, arguably a protective mechanism in this regard. Excessive amounts of ROS could cause additional mitochondrial damage, compromising residual respiration (Aragones et al. 2008) and ultimately end up in apoptosis (Denko 2008), and induce mutagenesis as well (Ishikawa et al. 2008). DJ-1, a redox sensitive protein that is activated upon oxidative stress, induces a protective and survival promoting activity via expression of antioxidant enzymes (Clements et al. 2006) and activation of the PI3/Akt pathway through inhibition of PTEN (Kim et al. 2005). Recently, DJ-1 has been shown to induce HIF-1α in tumor cell lines by activating the PI3/Akt/mTOR pathway, protecting the cell against hypoxic stress (Vasseur et al. 2009). Interestingly, DJ-1 was originally described as an oncogene, but loss of function mutations result in autosomal recessive hereditary Parkinson's disease (Bonifati et al. 2003). It is hypothesized that the cell survival dysfunction in both diseases is due to a lack versus an excess of antioxidant and HIF-mediated responses.

PHDs might arguably act as sensors of pH homeostasis as well, as increased levels of protons in low pH environments are expected to induce Fe^{2+} oxidation in the redox sensing core unit. By upregulating lactate and proton efflux mechanisms, a decrease in PHD activity prevents the detrimental effect of lowering pH levels on cell proliferation, and, on top of that, contributes to an acidic extracellular pH, which facilitates invasive tumor growth (Kroemer and Pouyssegur 2008; Swietach et al. 2007).

Finally, with 2-oxoglutarate as cosubstrate and succinate as end product, PHDs sense in a direct manner substrate availability for oxidative metabolism and biosynthesis. Interesting in this regard, several TCA enzymes are increasingly recognized as possible tumor suppressors (Fig. 6). Heterozygous germline mutation in

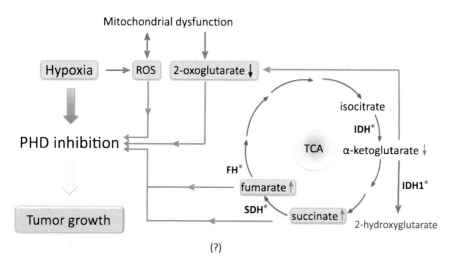

Fig. 6 PHDs as pivotal feedback molecules in cellular metabolism. PHDs not only sense oxygen availability but also respond to levels of oxygen radicals (ROS) and 2-oxoglutarate. In this way, PHDs act as sensors of cellular metabolism, orchestrating an adaptive metabolic reprogramming that allows cells to survive in conditions of reduced energy supply. In some tumors, genetic alterations in tumor suppressors are adopting this feedback mechanism. Loss of function mutations in FH and SDH (indicated by *asterisks*) result in increased levels of fumarate and succinate respectively, blocking the hydroxylation activity of PHD through "product inhibition". IDH mutations (indicated by *asterisks*) involve both loss and gain of function characteristics as reduced enzyme activity decreases the 2-oxoglutarate levels, whereas due to a pro-oncogenic function of the mutated IDH1, the putative tumorigenic metabolite 2-hydroxyglutarate is formed

the succinate dehydrogenase (SDH) gene and fumarate hydrase (FH) gene are associated with the development of pheochromocytoma or paraganglioma/leiomyosarcoma, respectively (Gottlieb and Tomlinson 2005). The molecular mechanism underlying this tumor susceptibility partly involves "product inhibition" by accumulation of succinate or the structural homologous fumarate (Isaacs et al. 2005; Selak et al. 2005). Succinate dehydrogenase is also functioning as complex II in the respiratory chain. Arguably, a dysfunctional electron transport could increase ROS generation, representing another mechanism of HIF upregulation (Guzy et al. 2008). Recent studies in glioma tumors revealed frequent heterozygous mutations in the gene, coding for cytosolic isocitrate dehydrogenase (IDH), which converts isocitrate to α-ketoglutarate (Yan et al. 2009a). Apart from catalyzing the conversion of α-ketoglutarate to the putative onco-metabolite R(-)-2-hydroxyglutarate (Dang et al. 2009), this mutation inhibits, via a dominant negative interaction, the residual wild type enzyme, thereby decreasing α-ketoglutarate levels and, indirectly through depletion of PHDs from their co-substrate, upregulating HIF-1α (Zhao et al. 2009). Intriguingly, an extended survival of patients carrying the IDH1 mutation was reported (Yan et al. 2009a).

In summary, PHD/HIF signaling directs a metabolic reprogramming that not only allows survival in the hostile tumor environment but also fuels malignant metastatic behavior. Several lines of experimental and clinical evidence point to a correlation between tumor metabolic features and poor outcome (Semenza 2009). As gatekeepers of energy and redox homeostasis, PHD sensors are a prototypic illustration of metabolism and tumorigenesis coming together.

5.2   Metabolism in Neurodegeneration: Food for Thought

Intermediates of the TCA cycle and free radicals affect the activity of PHDs, providing a feedback mechanism to link the efficacy and substrate availability of oxidative metabolism to the cellular hypoxia machinery. Mitochondrial dysfunction, impaired energy metabolism and oxidative stress represent familiar themes in neurodegeneration; these phenomena are interconnected in a self-perpetuating circle, which may culminate in neuronal death. Interestingly, genetic alterations of the mitochondrial enzymes SDH, FH and IDH that are involved in tumorigenesis are similarly implicated in neurodegenerative phenotypes. Bi-allelic germline mutations in SDHA and FH cause progressive infantile encephalopathies, known as Leigh syndrome (Bourgeron et al. 1995) and fumarate deficiency syndrome (Bourgeron et al. 1994), respectively. Moreover, 3-nitroproprionic, a succinate dehydrogenase inhibitor, induces selective striatal neuronal death and hence serves as a model for Huntington's disease (Garcia et al. 2002). Degeneration of photoreceptors in retinitis pigmentosa can result from loss of function mutations in the IDH3B gene (Hartong et al. 2008). Whereas in tumor biology a link between the mutated metabolic enzymes and the PHD/HIF pathway was established, it remains unclear whether this feedback mechanism contributes to neuronal death in these neurologic diseases as well. In the last section, we will discuss the involvement of deregulated energy metabolism at the crossroad of hypoxia signaling in ALS and AD.

5.2.1 Amyotrophic Lateral Sclerosis

With their axons extending more than one meter, motor neurons in humans are metabolically exceedingly demanding cells and therefore highly vulnerable to a disturbed energy homeostasis. Not surprisingly, both structural mitochondrial abnormalities and functional deficits with excessive ROS generation have been documented in ALS patients and rodent disease models (Hervias et al. 2006). Alternative evidence for deregulated metabolism comes from epidemiological data reporting glucose intolerance in a subset of ALS patients (Pradat et al. 2010). Rather unexpectedly was the observation of a hypermetabolic state with increased energy expenditure both in ALS patients as in the SOD1 mouse model (Bouteloup et al. 2009; Dupuis et al. 2004). Given the significant phenotypic improvement in SOD1 mice subjected to a high fat diet, this metabolic hypothesis warrants further mechanistic analysis. Whether the metabolic disturbances in ALS involve PHD signaling represents an exciting, unexplored area of research.

5.2.2 Alzheimer's Disease

Cerebral hypometabolism in AD patients has been a longstanding finding. FDG-PET imaging shows reduced glucose metabolism in specific brain circuitries, even in preclinical stages and presymptomatic subjects (Mosconi et al. 2008), suggesting a primary role of metabolic deficits in disease initiation and progression. Almost any level of glucose metabolism is affected in AD. First, the cellular glucose uptake is impaired. AD brains demonstrate reduced levels of GLUT1 and GLUT3, key neuronal glucose transporters (Liu et al. 2008). In animal models, decreased glucose metabolism is found to trigger tau hyperphosphorylation and consequently neurofibrillary tangle formation (Li et al. 2006). Even though glycolytic activity in AD is generally thought to be impaired (Bigl et al. 1999; Brooks et al. 2007), other groups report an enhanced activity of glycolytic enzymes in AD brains and mouse models (Soucek et al. 2003) and facilitated glycolytic flux in embryonic neuronal cultures from a triple transgenic AD mouse model (Yao et al. 2009), findings that will need to be reconciled better in the future. Impaired glucose oxidation is suggested by decreased expression and activity of mitochondrial enzymes and complexes in human autopsy samples (Blass et al. 2000; Brooks et al. 2007). These mitochondrial defects are found early in disease (Yao et al. 2009) and likely causally implicated, since AD cybrid cells (in which the mitochondria are replaced by mitochondria from AD patients) exhibit biochemical signs of mitochondrial dysfunction (Sheehan et al. 1997). Oxidative damage has been reported in AD brain (Andersen 2004), and is commonly designated as an important mediator of neuronal death. Some of these metabolic alterations may be related to Aβ, the pivotal neurotoxic player in AD. Indeed, Aβ reduces the glycolytic rate in astrocytes (Schubert et al. 2009), possibly via oxidation of glycolytic enzymes, and causes excessive ROS generation (Behl et al. 1994). Mitochondria are accumulating Aβ as well, where it directly interferes with mitochondrial function through interaction with complex III and IV (LaFerla et al. 2007).

An intriguing but controversial link between Aβ, glucose metabolism and HIF-1α-mediated hypoxia signaling has emerged over the past decade. A number of studies implicate HIF in an endogenous protective response against Aβ-induced toxicity. Increased glucose uptake, glycolytic flux and activity of the hexose monophosphate pathway, all induced by HIF-1α, have been considered as metabolic signature of Aβ resistant neural cell clones (Soucek et al. 2003). This metabolic program is argued to provide protection through generation of reducing equivalents via glycolysis and the hexose monophosphate shunt. Furthermore, HIF-1α induction reversed the Aβ-induced astrocyte activation and glycolytic reduction (Schubert et al. 2009). The question to what extent AD affects HIF-1α levels yields conflicting results: HIF-1α levels in autopsy specimens have been reported to be decreased (Liu et al. 2008) or upregulated (Soucek et al. 2003). Technical challenges to prevent rapid degradation of HIF proteins may complicate interpretation of some of these findings. Aβ has been shown to induce HIF-1α, which was hypothesized to occur through ROS generation (Acker and Acker 2004). Recent studies however refined this proposition and suggest that Aβ transiently inhibits proteosomal activity (resulting in HIF accumulation initially) but subsequently activates proteosomal degradation of HIF, offering an explanation for the observed reduced HIF levels (Guglielmotto et al. 2009; Schubert et al. 2009). The adverse effect of aging on the ability of a cell to induce the HIF pathway may further reduce neuroprotection in AD (Chavez and LaManna 2003).

All together, the precise role of metabolic alterations in the pathogenesis of neurodegeneration as well as the extent to which PHD signaling is involved remains largely obscure. In primary neuronal cultures at least, a metabolic adaptation involving upregulation of glycolytic enzymes and downregulation of oxidative metabolism is believed to contribute to their hypoxic resistance (Malthankar-Phatak et al. 2008). In neurodegenerative disorders, some evidence suggests a beneficial HIF-mediated metabolic reprogramming, as discussed above, although extensive work in the future will be needed to clarify the current controversies.

6 Conclusion

Cancer and neurodegeneration are two devastating age-associated diseases that cause a growing health burden in the aging population. Although these conditions are considered at the two ends of cell fate, they share a disturbed oxygen homeostasis in their pathogenesis. Whereas cancer cells adopt the hypoxic response to facilitate their growth, this protective mechanism is overwhelmed in neurodegenerative diseases. PHD signaling – at the crossroad between blood vessels and metabolism – involves both cell-autonomous activities as a crosstalk with the surrounding stroma. A better molecular characterization of the role of PHDs will likely lead to the development of effective therapies for both diseases.

Acknowledgments This work was supported by the K.U. Leuven, grant IUAP06/30 from the Belgian State, Federal Science Policy Office, Long term structural funding – Methusalem funding by the Flemish Government to PC; by a grants from the Fund for Scientific Research – Flanders (FWO G.0532.10 – G.0765.10 – 0652.08), by MND/A grant 07-6024, by FWO fellowship to A.Q.

References

Acker T, Acker H (2004) Cellular oxygen sensing need in CNS function: physiological and pathological implications. J Exp Biol 207:3171–3188

Acker T, Diez-Juan A, Aragones J, Tjwa M, Brusselmans K, Moons L, Fukumura D, Moreno-Murciano MP, Herbert JM, Burger A, Riedel J, Elvert G, Flamme I, Maxwell PH, Collen D, Dewerchin M, Jain RK, Plate KH, Carmeliet P (2005) Genetic evidence for a tumor suppressor role of HIF-2alpha. Cancer Cell 8:131–141

Andersen JK (2004) Oxidative stress in neurodegeneration: cause or consequence? Nat Rev Neurosci 10 Suppl:S18–S25

Aragones J, Schneider M, Van Geyte K, Fraisl P, Dresselaers T, Mazzone M, Dirkx R, Zacchigna S, Lemieux H, Jeoung NH, Lambrechts D, Bishop T, Lafuste P, Diez-Juan A, Harten SK, Van Noten P, De Bock K, Willam C, Tjwa M, Grosfeld A, Navet R, Moons L, Vandendriessche T, Deroose C, Wijeyekoon B, Nuyts J, Jordan B, Silasi-Mansat R, Lupu F, Dewerchin M, Pugh C, Salmon P, Mortelmans L, Gallez B, Gorus F, Buyse J, Sluse F, Harris RA, Gnaiger E, Hespel P, Van Hecke P, Schuit F, Van Veldhoven P, Ratcliffe P, Baes M, Maxwell P, Carmeliet P (2008) Deficiency or inhibition of oxygen sensor Phd1 induces hypoxia tolerance by reprogramming basal metabolism. Nat Genet 40:170–180

Aragones J, Fraisl P, Baes M, Carmeliet P (2009) Oxygen sensors at the crossroad of metabolism. Cell Metab 9:11–22

Azzouz M, Ralph GS, Storkebaum E, Walmsley LE, Mitrophanous KA, Kingsman SM, Carmeliet P, Mazarakis ND (2004) VEGF delivery with retrogradely transported lentivector prolongs survival in a mouse ALS model. Nature 429:413–417

Baranova O, Miranda LF, Pichiule P, Dragatsis I, Johnson RS, Chavez JC (2007) Neuron-specific inactivation of the hypoxia inducible factor 1 alpha increases brain injury in a mouse model of transient focal cerebral ischemia. J Neurosci 27:6320–6332

Behl C, Davis JB, Lesley R, Schubert D (1994) Hydrogen peroxide mediates amyloid beta protein toxicity. Cell 77:817–827

Bell RD, Zlokovic BV (2009) Neurovascular mechanisms and blood-brain barrier disorder in Alzheimer's disease. Acta Neuropathol 118:103–113

Bell RD, Deane R, Chow N, Long X, Sagare A, Singh I, Streb JW, Guo H, Rubio A, Van Nostrand W, Miano JM, Zlokovic BV (2009) SRF and myocardin regulate LRP-mediated amyloid-beta clearance in brain vascular cells. Nat Cell Biol 11:143–153

Ben-Shachar D, Eshel G, Finberg JP, Youdim MB (1991) The iron chelator desferrioxamine (Desferal) retards 6-hydroxydopamine-induced degeneration of nigrostriatal dopamine neurons. J Neurochem 56:1441–1444

Bergers G, Hanahan D (2008) Modes of resistance to anti-angiogenic therapy. Nat Rev Cancer 8:592–603

Bernaudin M, Tang Y, Reilly M, Petit E, Sharp FR (2002) Brain genomic response following hypoxia and re-oxygenation in the neonatal rat. Identification of genes that might contribute to hypoxia-induced ischemic tolerance. J Biol Chem 277:39728–39738

Bertout JA, Patel SA, Simon MC (2008) The impact of O2 availability on human cancer. Nat Rev Cancer 8:967–975

Bigl M, Bruckner MK, Arendt T, Bigl V, Eschrich K (1999) Activities of key glycolytic enzymes in the brains of patients with Alzheimer's disease. J Neural Transm 106:499–511

Blass JP, Sheu RK, Gibson GE (2000) Inherent abnormalities in energy metabolism in Alzheimer disease. Interaction with cerebrovascular compromise. Ann N Y Acad Sci 903:204–221

Bonifati V, Rizzu P, van Baren MJ, Schaap O, Breedveld GJ, Krieger E, Dekker MC, Squitieri F, Ibanez P, Joosse M, van Dongen JW, Vanacore N, van Swieten JC, Brice A, Meco G, van Duijn CM, Oostra BA, Heutink P (2003) Mutations in the DJ-1 gene associated with autosomal recessive early-onset parkinsonism. Science 299:256–259

Bookheimer SY, Strojwas MH, Cohen MS, Saunders AM, Pericak-Vance MA, Mazziotta JC, Small GW (2000) Patterns of brain activation in people at risk for Alzheimer's disease. N Engl J Med 343:450–456

Bourgeron T, Chretien D, Poggi-Bach J, Doonan S, Rabier D, Letouze P, Munnich A, Rotig A, Landrieu P, Rustin P (1994) Mutation of the fumarase gene in two siblings with progressive encephalopathy and fumarase deficiency. J Clin Invest 93:2514–2518

Bourgeron T, Rustin P, Chretien D, Birch-Machin M, Bourgeois M, Viegas-Pequignot E, Munnich A, Rotig A (1995) Mutation of a nuclear succinate dehydrogenase gene results in mitochondrial respiratory chain deficiency. Nat Genet 11:144–149

Bouteloup C, Desport JC, Clavelou P, Guy N, Derumeaux-Burel H, Ferrier A, Couratier P (2009) Hypermetabolism in ALS patients: an early and persistent phenomenon. J Neurol 256: 1236–1242

Brines M, Cerami A (2005) Emerging biological roles for erythropoietin in the nervous system. Nat Rev Neurosci 6:484–494

Brokers N, Le-Huu S, Vogel S, Hagos Y, Katschinski DM, Kleinschmidt M (2010) Increased chemoresistance induced by inhibition of HIF-prolyl-hydroxylase domain enzymes. Cancer Sci 101(1):129–136

Brooks WM, Lynch PJ, Ingle CC, Hatton A, Emson PC, Faull RL, Starkey MP (2007) Gene expression profiles of metabolic enzyme transcripts in Alzheimer's disease. Brain Res 1127:127–135

Brown JM, Wilson WR (2004) Exploiting tumour hypoxia in cancer treatment. Nat Rev Cancer 4:437–447

Calvisi DF, Ladu S, Gorden A, Farina M, Lee JS, Conner EA, Schroeder I, Factor VM, Thorgeirsson SS (2007) Mechanistic and prognostic significance of aberrant methylation in the molecular pathogenesis of human hepatocellular carcinoma. J Clin Invest 117:2713–2722

Carmeliet P (2005) Angiogenesis in life, disease and medicine. Nature 438:932–936

Carmeliet P, Jain RK (2000) Angiogenesis in cancer and other diseases. Nature 407:249–257

Carmeliet P, Dor Y, Herbert JM, Fukumura D, Brusselmans K, Dewerchin M, Neeman M, Bono F, Abramovitch R, Maxwell P, Koch CJ, Ratcliffe P, Moons L, Jain RK, Collen D, Keshert E (1998) Role of HIF-1alpha in hypoxia-mediated apoptosis, cell proliferation and tumour angiogenesis. Nature 394:485–490

Casanovas O, Hicklin DJ, Bergers G, Hanahan D (2005) Drug resistance by evasion of antiangiogenic targeting of VEGF signaling in late-stage pancreatic islet tumors. Cancer Cell 8:299–309

Chan DA, Sutphin PD, Denko NC, Giaccia AJ (2002) Role of prolyl hydroxylation in oncogenically stabilized hypoxia-inducible factor-1alpha. J Biol Chem 277:40112–40117

Chan DA, Kawahara TL, Sutphin PD, Chang HY, Chi JT, Giaccia AJ (2009) Tumor vasculature is regulated by PHD2-mediated angiogenesis and bone marrow-derived cell recruitment. Cancer Cell 15:527–538

Chavez JC, LaManna JC (2003) Hypoxia-inducible factor-1alpha accumulation in the rat brain in response to hypoxia and ischemia is attenuated during aging. Adv Exp Med Biol 510:337–341

Chen W, Jadhav V, Tang J, Zhang JH (2008) HIF-1alpha inhibition ameliorates neonatal brain injury in a rat pup hypoxic-ischemic model. Neurobiol Dis 31:433–441

Chen C, Hu Q, Yan J, Yang X, Shi X, Lei J, Chen L, Huang H, Han J, Zhang JH, Zhou C (2009a) Early inhibition of HIF-1alpha with small interfering RNA reduces ischemic-reperfused brain injury in rats. Neurobiol Dis 33:509–517

Chen W, Ostrowski RP, Obenaus A, Zhang JH (2009b) Prodeath or prosurvival: two facets of hypoxia inducible factor-1 in perinatal brain injury. Exp Neurol 216(1):7–15

Chiappelli M, Borroni B, Archetti S, Calabrese E, Corsi MM, Franceschi M, Padovani A, Licastro F (2006) VEGF gene and phenotype relation with Alzheimer's disease and mild cognitive impairment. Rejuvenation Res 9:485–493

Chow N, Bell RD, Deane R, Streb JW, Chen J, Brooks A, Van Nostrand W, Miano JM, Zlokovic BV (2007) Serum response factor and myocardin mediate arterial hypercontractility and cerebral blood flow dysregulation in Alzheimer's phenotype. Proc Natl Acad Sci USA 104:823–828

Clements CM, McNally RS, Conti BJ, Mak TW, Ting JP (2006) DJ-1, a cancer- and Parkinson's disease-associated protein, stabilizes the antioxidant transcriptional master regulator Nrf2. Proc Natl Acad Sci USA 103:15091–15096

Coleman ML, McDonough MA, Hewitson KS, Coles C, Mecinovic J, Edelmann M, Cook KM, Cockman ME, Lancaster DE, Kessler BM, Oldham NJ, Ratcliffe PJ, Schofield CJ (2007) Asparaginyl hydroxylation of the Notch ankyrin repeat domain by factor inhibiting hypoxia-inducible factor. J Biol Chem 282:24027–24038

Correia SC, Moreira PI (2010) Hypoxia-inducible factor 1: a new hope to counteract neurodegeneration? J Neurochem 112(1):1–12

Couvelard A, Deschamps L, Rebours V, Sauvanet A, Gatter K, Pezzella F, Ruszniewski P, Bedossa P (2008) Overexpression of the oxygen sensors PHD-1, PHD-2, PHD-3, and FIH Is associated with tumor aggressiveness in pancreatic endocrine tumors. Clin Cancer Res 14:6634–6639

Crapper McLachlan DR, Dalton AJ, Kruck TP, Bell MY, Smith WL, Kalow W, Andrews DF (1991) Intramuscular desferrioxamine in patients with Alzheimer's disease. Lancet 337: 1304–1308

Crawford Y, Ferrara N (2009) Tumor and stromal pathways mediating refractoriness/resistance to anti-angiogenic therapies. Trends Pharmacol Sci 30:624–630

Crawford Y, Kasman I, Yu L, Zhong C, Wu X, Modrusan Z, Kaminker J, Ferrara N (2009) PDGF-C mediates the angiogenic and tumorigenic properties of fibroblasts associated with tumors refractory to anti-VEGF treatment. Cancer Cell 15:21–34

Cummins EP, Berra E, Comerford KM, Ginouves A, Fitzgerald KT, Seeballuck F, Godson C, Nielsen JE, Moynagh P, Pouyssegur J, Taylor CT (2006) Prolyl hydroxylase-1 negatively regulates IkappaB kinase-beta, giving insight into hypoxia-induced NFkappaB activity. Proc Natl Acad Sci USA 103:18154–18159

Dang L, White DW, Gross S, Bennett BD, Bittinger MA, Driggers EM, Fantin VR, Jang HG, Jin S, Keenan MC, Marks KM, Prins RM, Ward PS, Yen KE, Liau LM, Rabinowitz JD, Cantley LC, Thompson CB, Vander Heiden MG, Su SM (2009) Cancer-associated IDH1 mutations produce 2-hydroxyglutarate. Nature 462:739–744

de la Torre JC (2004) Is Alzheimer's disease a neurodegenerative or a vascular disorder? Data, dogma, and dialectics. Lancet Neurol 3:184–190

Deberardinis RJ, Sayed N, Ditsworth D, Thompson CB (2008) Brick by brick: metabolism and tumor cell growth. Curr Opin Genet Dev 18:54–61

Denko NC (2008) Hypoxia, HIF1 and glucose metabolism in the solid tumour. Nat Rev Cancer 8:705–713

Devos D, Moreau C, Lassalle P, Perez T, De Seze J, Brunaud-Danel V, Destee A, Tonnel AB, Just N (2004) Low levels of the vascular endothelial growth factor in CSF from early ALS patients. Neurology 62:2127–2129

Dewhirst MW, Cao Y, Moeller B (2008) Cycling hypoxia and free radicals regulate angiogenesis and radiotherapy response. Nat Rev Cancer 8:425–437

Dirnagl U, Becker K, Meisel A (2009) Preconditioning and tolerance against cerebral ischaemia: from experimental strategies to clinical use. Lancet Neurol 8:398–412

Du R, Lu KV, Petritsch C, Liu P, Ganss R, Passegue E, Song H, Vandenberg S, Johnson RS, Werb Z, Bergers G (2008) HIF1alpha induces the recruitment of bone marrow-derived vascular modulatory cells to regulate tumor angiogenesis and invasion. Cancer Cell 13:206–220

Dupuis L, Oudart H, Rene F, Gonzalez de Aguilar JL, Loeffler JP (2004) Evidence for defective energy homeostasis in amyotrophic lateral sclerosis: benefit of a high-energy diet in a transgenic mouse model. Proc Natl Acad Sci USA 101:11159–11164

Ebos JM, Lee CR, Cruz-Munoz W, Bjarnason GA, Christensen JG, Kerbel RS (2009) Accelerated metastasis after short-term treatment with a potent inhibitor of tumor angiogenesis. Cancer Cell 15:232–239

Eckle T, Kohler D, Lehmann R, El Kasmi K, Eltzschig HK (2008) Hypoxia-inducible factor-1 is central to cardioprotection: a new paradigm for ischemic preconditioning. Circulation 118:166–175

Ellis LM, Hicklin DJ (2008) VEGF-targeted therapy: mechanisms of anti-tumour activity. Nat Rev Cancer 8:579–591

Fernando NT, Koch M, Rothrock C, Gollogly LK, D'Amore PA, Ryeom S, Yoon SS (2008) Tumor escape from endogenous, extracellular matrix-associated angiogenesis inhibitors by up-regulation of multiple proangiogenic factors. Clin Cancer Res 14:1529–1539

Fischer C, Mazzone M, Jonckx B, Carmeliet P (2008) FLT1 and its ligands VEGFB and PlGF: drug targets for anti-angiogenic therapy? Nat Rev Cancer 8:942–956

Folkman J (1971) Tumor angiogenesis: therapeutic implications. N Engl J Med 285:1182–1186

Folkman J (2007) Angiogenesis: an organizing principle for drug discovery? Nat Rev Drug Discov 6:273–286

Fraisl P, Aragones J, Carmeliet P (2009) Inhibition of oxygen sensors as a therapeutic strategy for ischaemic and inflammatory disease. Nat Rev Drug Discov 8:139–152

Franco M, Man S, Chen L, Emmenegger U, Shaked Y, Cheung AM, Brown AS, Hicklin DJ, Foster FS, Kerbel RS (2006) Targeted anti-vascular endothelial growth factor receptor-2 therapy leads to short-term and long-term impairment of vascular function and increase in tumor hypoxia. Cancer Res 66:3639–3648

Freret T, Valable S, Chazalviel L, Saulnier R, Mackenzie ET, Petit E, Bernaudin M, Boulouard M, Schumann-Bard P (2006) Delayed administration of deferoxamine reduces brain damage and promotes functional recovery after transient focal cerebral ischemia in the rat. Eur J Neurosci 23:1757–1765

Garbuzova-Davis S, Saporta S, Haller E, Kolomey I, Bennett SP, Potter H, Sanberg PR (2007) Evidence of compromised blood-spinal cord barrier in early and late symptomatic SOD1 mice modeling ALS. PLoS One 2:e1205

Garcia M, Vanhoutte P, Pages C, Besson MJ, Brouillet E, Caboche J (2002) The mitochondrial toxin 3-nitropropionic acid induces striatal neurodegeneration via a c-Jun N-terminal kinase/c-Jun module. J Neurosci 22:2174–2184

Gatenby RA, Gillies RJ (2004) Why do cancers have high aerobic glycolysis? Nat Rev Cancer 4:891–899

Gerald D, Berra E, Frapart YM, Chan DA, Giaccia AJ, Mansuy D, Pouyssegur J, Yaniv M, Mechta-Grigoriou F (2004) JunD reduces tumor angiogenesis by protecting cells from oxidative stress. Cell 118:781–794

Gidday JM (2006) Cerebral preconditioning and ischaemic tolerance. Nat Rev Neurosci 7:437–448

Gottlieb E, Tomlinson IP (2005) Mitochondrial tumour suppressors: a genetic and biochemical update. Nat Rev Cancer 5:857–866

Graeber TG, Osmanian C, Jacks T, Housman DE, Koch CJ, Lowe SW, Giaccia AJ (1996) Hypoxia-mediated selection of cells with diminished apoptotic potential in solid tumours. Nature 379:88–91

Greenberg SM, Gurol ME, Rosand J, Smith EE (2004) Amyloid angiopathy-related vascular cognitive impairment. Stroke 35:2616–2619

Greenberg JI, Shields DJ, Barillas SG, Acevedo LM, Murphy E, Huang J, Scheppke L, Stockmann C, Johnson RS, Angle N, Cheresh DA (2008) A role for VEGF as a negative regulator of pericyte function and vessel maturation. Nature 456:809–813

Greenway MJ, Andersen PM, Russ C, Ennis S, Cashman S, Donaghy C, Patterson V, Swingler R, Kieran D, Prehn J, Morrison KE, Green A, Acharya KR, Brown RH Jr, Hardiman O (2006) ANG mutations segregate with familial and 'sporadic' amyotrophic lateral sclerosis. Nat Genet 38:411–413

Guglielmotto M, Aragno M, Autelli R, Giliberto L, Novo E, Colombatto S, Danni O, Parola M, Smith MA, Perry G, Tamagno E, Tabaton M (2009) The up-regulation of BACE1 mediated by hypoxia and ischemic injury: role of oxidative stress and HIF1alpha. J Neurochem 108: 1045–1056

Guzy RD, Schumacker PT (2006) Oxygen sensing by mitochondria at complex III: the paradox of increased reactive oxygen species during hypoxia. Exp Physiol 91:807–819

Guzy RD, Sharma B, Bell E, Chandel NS, Schumacker PT (2008) Loss of the SdhB, but Not the SdhA, subunit of complex II triggers reactive oxygen species-dependent hypoxia-inducible factor activation and tumorigenesis. Mol Cell Biol 28:718–731

Haass C, Selkoe DJ (2007) Soluble protein oligomers in neurodegeneration: lessons from the Alzheimer's amyloid beta-peptide. Nat Rev Mol Cell Biol 8:101–112

Hamrick SE, McQuillen PS, Jiang X, Mu D, Madan A, Ferriero DM (2005) A role for hypoxia-inducible factor-1alpha in desferoxamine neuroprotection. Neurosci Lett 379:96–100

Hanahan D, Folkman J (1996) Patterns and emerging mechanisms of the angiogenic switch during tumorigenesis. Cell 86:353–364

Hanahan D, Weinberg RA (2000) The hallmarks of cancer. Cell 100:57–70

Harris AL (2002) Hypoxia – a key regulatory factor in tumour growth. Nat Rev Cancer 2:38–47

Harten SK, Ashcroft M, Maxwell PH (2010) Prolyl hydroxylase domain inhibitors: a route to HIF activation and neuroprotection. Antioxid Redox Signal 12(4):459–480

Hartong DT, Dange M, McGee TL, Berson EL, Dryja TP, Colman RF (2008) Insights from retinitis pigmentosa into the roles of isocitrate dehydrogenases in the Krebs cycle. Nat Genet 40:1230–1234

Helmlinger G, Yuan F, Dellian M, Jain RK (1997) Interstitial pH and pO2 gradients in solid tumors *in vivo*: high-resolution measurements reveal a lack of correlation. Nat Med 3:177–182

Helton R, Cui J, Scheel JR, Ellison JA, Ames C, Gibson C, Blouw B, Ouyang L, Dragatsis I, Zeitlin S, Johnson RS, Lipton SA, Barlow C (2005) Brain-specific knock-out of hypoxia-inducible factor-1alpha reduces rather than increases hypoxic-ischemic damage. J Neurosci 25:4099–4107

Henze AT, Riedel J, Diem T, Wenner J, Flamme I, Pouysseggur J, Plate KH, Acker T (2010) Prolyl hydroxylases 2 and 3 act in gliomas as protective negative feedback regulators of hypoxia-inducible factors. Cancer Res 70(1):357–366

Hervias I, Beal MF, Manfredi G (2006) Mitochondrial dysfunction and amyotrophic lateral sclerosis. Muscle Nerve 33:598–608

Hyvarinen J, Hassinen IE, Sormunen R, Maki JM, Kivirikko KI, Koivunen P, Myllyharju J (2010) Hearts of hypoxia-inducible factor prolyl 4-hydroxylase-2 hypomorphic mice show protection against acute ischemia-reperfusion injury. J Biol Chem 285(18):13646–13657

Iadecola C (2004) Neurovascular regulation in the normal brain and in Alzheimer's disease. Nat Rev Neurosci 5:347–360

Iadecola C, Nedergaard M (2007) Glial regulation of the cerebral microvasculature. Nat Neurosci 10:1369–1376

Iadecola C, Zhang F, Niwa K, Eckman C, Turner SK, Fischer E, Younkin S, Borchelt DR, Hsiao KK, Carlson GA (1999) SOD1 rescues cerebral endothelial dysfunction in mice overexpressing amyloid precursor protein. Nat Neurosci 2:157–161

Isaacs JS, Jung YJ, Mole DR, Lee S, Torres-Cabala C, Chung YL, Merino M, Trepel J, Zbar B, Toro J, Ratcliffe PJ, Linehan WM, Neckers L (2005) HIF overexpression correlates with biallelic loss of fumarate hydratase in renal cancer: novel role of fumarate in regulation of HIF stability. Cancer Cell 8:143–153

Ishikawa K, Takenaga K, Akimoto M, Koshikawa N, Yamaguchi A, Imanishi H, Nakada K, Honma Y, Hayashi J (2008) ROS-generating mitochondrial DNA mutations can regulate tumor cell metastasis. Science 320:661–664

Jain RK (2005) Normalization of tumor vasculature: an emerging concept in antiangiogenic therapy. Science 307:58–62

Jokilehto T, Rantanen K, Luukkaa M, Heikkinen P, Grenman R, Minn H, Kronqvist P, Jaakkola PM (2006) Overexpression and nuclear translocation of hypoxia-inducible factor prolyl hydroxylase PHD2 in head and neck squamous cell carcinoma is associated with tumor aggressiveness. Clin Cancer Res 12:1080–1087

Jones RG, Thompson CB (2009) Tumor suppressors and cell metabolism: a recipe for cancer growth. Genes Dev 23:537–548

Just N, Moreau C, Lassalle P, Gosset P, Perez T, Brunaud-Danel V, Wallaert B, Destee A, Defebvre L, Tonnel AB, Devos D (2007) High erythropoietin and low vascular endothelial growth factor levels in cerebrospinal fluid from hypoxemic ALS patients suggest an abnormal response to hypoxia. Neuromuscul Disord 17:169–173

Kaelin WG Jr (2008) The von Hippel-Lindau tumour suppressor protein: O2 sensing and cancer. Nat Rev Cancer 8:865–873

Kaelin WG Jr, Ratcliffe PJ (2008) Oxygen sensing by metazoans: the central role of the HIF hydroxylase pathway. Mol Cell 30:393–402

Kato H, Inoue T, Asanoma K, Nishimura C, Matsuda T, Wake N (2006) Induction of human endometrial cancer cell senescence through modulation of HIF-1alpha activity by EGLN1. Int J Cancer 118:1144–1153

Kim RH, Peters M, Jang Y, Shi W, Pintilie M, Fletcher GC, DeLuca C, Liepa J, Zhou L, Snow B, Binari RC, Manoukian AS, Bray MR, Liu FF, Tsao MS, Mak TW (2005) DJ-1, a novel regulator of the tumor suppressor PTEN. Cancer Cell 7:263–273

Knowles HJ, Raval RR, Harris AL, Ratcliffe PJ (2003) Effect of ascorbate on the activity of hypoxia-inducible factor in cancer cells. Cancer Res 63:1764–1768

Kroemer G, Pouyssegur J (2008) Tumor cell metabolism: cancer's Achilles' heel. Cancer Cell 13:472–482

Kulkarni AC, Kuppusamy P, Parinandi N (2007) Oxygen, the lead actor in the pathophysiologic drama: enactment of the trinity of normoxia, hypoxia, and hyperoxia in disease and therapy. Antioxid Redox Signal 9:1717–1730

Kupershmidt L, Weinreb O, Amit T, Mandel S, Carri MT, Youdim MB (2009) Neuroprotective and neuritogenic activities of novel multimodal iron-chelating drugs in motor-neuron-like NSC-34 cells and transgenic mouse model of amyotrophic lateral sclerosis. FASEB J 23:3766–3779

LaFerla FM, Green KN, Oddo S (2007) Intracellular amyloid-beta in Alzheimer's disease. Nat Rev Neurosci 8:499–509

Lambrechts D, Poesen K, Fernandez-Santiago R, Al-Chalabi A, Del Bo R, Van Vught PW, Khan S, Marklund SL, Brockington A, van Marion I, Anneser J, Shaw C, Ludolph AC, Leigh NP, Comi GP, Gasser T, Shaw PJ, Morrison KE, Andersen PM, Van den Berg LH, Thijs V, Siddique T, Robberecht W, Carmeliet P (2009) Meta-analysis of vascular endothelial growth factor variations in amyotrophic lateral sclerosis: increased susceptibility in male carriers of the -2578AA genotype. J Med Genet 46:840–846

Lan J, Jiang DH (1997) Desferrioxamine and vitamin E protect against iron and MPTP-induced neurodegeneration in mice. J Neural Transm 104:469–481

Lee KA, Lynd JD, O'Reilly S, Kiupel M, McCormick JJ, LaPres JJ (2008) The biphasic role of the hypoxia-inducible factor prolyl-4-hydroxylase, PHD2, in modulating tumor-forming potential. Mol Cancer Res 6:829–842

Lendahl U, Lee KL, Yang H, Poellinger L (2009) Generating specificity and diversity in the transcriptional response to hypoxia. Nat Rev Genet 10:821–832

Li X, Lu F, Wang JZ, Gong CX (2006) Concurrent alterations of O-GlcNAcylation and phosphorylation of tau in mouse brains during fasting. Eur J Neurosci 23:2078–2086

Li YX, Ding SJ, Xiao L, Guo W, Zhan Q (2008) Desferoxamine preconditioning protects against cerebral ischemia in rats by inducing expressions of hypoxia inducible factor 1 alpha and erythropoietin. Neurosci Bull 24:89–95

Li L, Zhang X, Yang D, Luo G, Chen S, Le W (2009) Hypoxia increases Abeta generation by altering beta- and gamma-cleavage of APP. Neurobiol Aging 30:1091–1098

Liu Y, Liu F, Iqbal K, Grundke-Iqbal I, Gong CX (2008) Decreased glucose transporters correlate to abnormal hyperphosphorylation of tau in Alzheimer disease. FEBS Lett 582:359–364

Loges S, Mazzone M, Hohensinner P, Carmeliet P (2009) Silencing or fueling metastasis with VEGF inhibitors: antiangiogenesis revisited. Cancer Cell 15:167–170

Luukkaa M, Jokilehto T, Kronqvist P, Vahlberg T, Grenman R, Jaakkola P, Minn H (2009) Expression of the cellular oxygen sensor PHD2 (EGLN-1) predicts radiation sensitivity in squamous cell cancer of the head and neck. Int J Radiat Biol 85(10):900–908

Malthankar-Phatak GH, Patel AB, Xia Y, Hong S, Chowdhury GM, Behar KL, Orina IA, Lai JC (2008) Effects of continuous hypoxia on energy metabolism in cultured cerebro-cortical neurons. Brain Res 1229:147–154

Marxsen JH, Stengel P, Doege K, Heikkinen P, Jokilehto T, Wagner T, Jelkmann W, Jaakkola P, Metzen E (2004) Hypoxia-inducible factor-1 (HIF-1) promotes its degradation by induction of HIF-alpha-prolyl-4-hydroxylases. Biochem J 381:761–767

Matsumoto K, Obara N, Ema M, Horie M, Naka A, Takahashi S, Imagawa S (2009) Antitumor effects of 2-oxoglutarate through inhibition of angiogenesis in a murine tumor model. Cancer Sci 100:1639–1647

Mazzone M, Dettori D, Leite de Oliveira R, Loges S, Schmidt T, Jonckx B, Tian YM, Lanahan AA, Pollard P, Ruiz de Almodovar C, De Smet F, Vinckier S, Aragones J, Debackere K, Luttun A, Wyns S, Jordan B, Pisacane A, Gallez B, Lampugnani MG, Dejana E, Simons M, Ratcliffe P, Maxwell P, Carmeliet P (2009) Heterozygous deficiency of PHD2 restores tumor oxygenation and inhibits metastasis via endothelial normalization. Cell 136:839–851

Metzen E, Berchner-Pfannschmidt U, Stengel P, Marxsen JH, Stolze I, Klinger M, Huang WQ, Wotzlaw C, Hellwig-Burgel T, Jelkmann W, Acker H, Fandrey J (2003) Intracellular localisation of human HIF-1 alpha hydroxylases: implications for oxygen sensing. J Cell Sci 116:1319–1326

Micieli JA, Santiago P, Brent MH (2009) Third nerve palsy following intravitreal anti-VEGF therapy for bilateral neovascular age-related macular degeneration. Acta Ophthalmol

Mikhaylova O, Ignacak ML, Barankiewicz TJ, Harbaugh SV, Yi Y, Maxwell PH, Schneider M, Van Geyte K, Carmeliet P, Revelo MP, Wyder M, Greis KD, Meller J, Czyzyk-Krzeska MF (2008) The von Hippel-Lindau tumor suppressor protein and Egl-9-Type proline hydroxylases regulate the large subunit of RNA polymerase II in response to oxidative stress. Mol Cell Biol 28:2701–2717

Moreau C, Devos D, Brunaud-Danel V, Defebvre L, Perez T, Destee A, Tonnel AB, Lassalle P, Just N (2006) Paradoxical response of VEGF expression to hypoxia in CSF of patients with ALS. J Neurol Neurosurg Psychiatry 77:255–257

Moreau C, Gosset P, Brunaud-Danel V, Lassalle P, Degonne B, Destee A, Defebvre L, Devos D (2009) CSF profiles of angiogenic and inflammatory factors depend on the respiratory status of ALS patients. Amyotroph Lateral Scler 10:175–181

Mosconi L, Pupi A, De Leon MJ (2008) Brain glucose hypometabolism and oxidative stress in preclinical Alzheimer's disease. Ann N Y Acad Sci 1147:180–195

Murakami T, Ilieva H, Shiote M, Nagata T, Nagano I, Shoji M, Abe K (2003) Hypoxic induction of vascular endothelial growth factor is selectively impaired in mice carrying the mutant SOD1 gene. Brain Res 989:231–237

Narayana A, Kelly P, Golfinos J, Parker E, Johnson G, Knopp E, Zagzag D, Fischer I, Raza S, Medabalmi P, Eagan P, Gruber ML (2009) Antiangiogenic therapy using bevacizumab in recurrent high-grade glioma: impact on local control and patient survival. J Neurosurg 110:173–180

Nishijima K, Ng YS, Zhong L, Bradley J, Schubert W, Jo N, Akita J, Samuelsson SJ, Robinson GS, Adamis AP, Shima DT (2007) Vascular endothelial growth factor-A is a survival factor for retinal neurons and a critical neuroprotectant during the adaptive response to ischemic injury. Am J Pathol 171:53–67

Niwa K, Younkin L, Ebeling C, Turner SK, Westaway D, Younkin S, Ashe KH, Carlson GA, Iadecola C (2000) Abeta 1-40-related reduction in functional hyperemia in mouse neocortex during somatosensory activation. Proc Natl Acad Sci USA 97:9735–9740

Norden AD, Young GS, Setayesh K, Muzikansky A, Klufas R, Ross GL, Ciampa AS, Ebeling LG, Levy B, Drappatz J, Kesari S, Wen PY (2008) Bevacizumab for recurrent malignant gliomas: efficacy, toxicity, and patterns of recurrence. Neurology 70:779–787

Oosthuyse B, Moons L, Storkebaum E, Beck H, Nuyens D, Brusselmans K, Van Dorpe J, Hellings P, Gorselink M, Heymans S, Theilmeier G, Dewerchin M, Laudenbach V, Vermylen P, Raat H, Acker T, Vleminckx V, Van Den Bosch L, Cashman N, Fujisawa H, Drost MR, Sciot R, Bruyninckx F, Hicklin DJ, Ince C, Gressens P, Lupu F, Plate KH, Robberecht W, Herbert JM, Collen D, Carmeliet P (2001) Deletion of the hypoxia-response element in the vascular endothelial growth factor promoter causes motor neuron degeneration. Nat Genet 28: 131–138

Paez-Ribes M, Allen E, Hudock J, Takeda T, Okuyama H, Vinals F, Inoue M, Bergers G, Hanahan D, Casanovas O (2009) Antiangiogenic therapy elicits malignant progression of tumors to increased local invasion and distant metastasis. Cancer Cell 15:220–231

Paris D, Patel N, DelleDonne A, Quadros A, Smeed R, Mullan M (2004a) Impaired angiogenesis in a transgenic mouse model of cerebral amyloidosis. Neurosci Lett 366:80–85

Paris D, Townsend K, Quadros A, Humphrey J, Sun J, Brem S, Wotoczek-Obadia M, DelleDonne A, Patel N, Obregon DF, Crescentini R, Abdullah L, Coppola D, Rojiani AM, Crawford F, Sebti SM, Mullan M (2004b) Inhibition of angiogenesis by Abeta peptides. Angiogenesis 7:75–85

Park L, Anrather J, Zhou P, Frys K, Pitstick R, Younkin S, Carlson GA, Iadecola C (2005) NADPH-oxidase-derived reactive oxygen species mediate the cerebrovascular dysfunction induced by the amyloid beta peptide. J Neurosci 25:1769–1777

Patel NS, Mathura VS, Bachmeier C, Beaulieu-Abdelahad D, Laporte V, Weeks O, Mullan M, Paris D (2010) Alzheimer's beta-amyloid peptide blocks vascular endothelial growth factor mediated signaling via direct interaction with VEGFR-2. J Neurochem 112(1):66–76

Pouyssegur J, Dayan F, Mazure NM (2006) Hypoxia signalling in cancer and approaches to enforce tumour regression. Nature 441:437–443

Pradat PF, Bruneteau G, Gordon PH, Dupuis L, Bonnefont-Rousselot D, Simon D, Salachas F, Corcia P, Frochot V, Lacorte JM, Jardel C, Coussieu C, Forestier NL, Lacomblez L, Loeffler JP, Meininger V (2010) Impaired glucose tolerance in patients with amyotrophic lateral sclerosis. Amyotroph Lateral Scler 11(1–2):166–171

Qi J, Nakayama K, Gaitonde S, Goydos JS, Krajewski S, Eroshkin A, Bar-Sagi D, Bowtell D, Ronai Z (2008) The ubiquitin ligase Siah2 regulates tumorigenesis and metastasis by HIF-dependent and -independent pathways. Proc Natl Acad Sci USA 105:16713–16718

Rankin EB, Giaccia AJ (2008) The role of hypoxia-inducible factors in tumorigenesis. Cell Death Differ 15:678–685

Ratan RR, Siddiq A, Aminova L, Langley B, McConoughey S, Karpisheva K, Lee HH, Carmichael T, Kornblum H, Coppola G, Geschwind DH, Hoke A, Smirnova N, Rink C, Roy S, Sen C, Beattie MS, Hart RP, Grumet M, Sun D, Freeman RS, Semenza GL, Gazaryan I (2008) Small molecule activation of adaptive gene expression: tilorone or its analogs are novel potent activators of hypoxia inducible factor-1 that provide prophylaxis against stroke and spinal cord injury. Ann N Y Acad Sci 1147:383–394

Rieger J, Bahr O, Muller K, Franz K, Steinbach J, Hattingen E (2009) Bevacizumab-induced diffusion-restricted lesions in malignant glioma patients. J Neurooncol

Rowland LP, Shneider NA (2001) Amyotrophic lateral sclerosis. N Engl J Med 344:1688–1700

Ruiz de Almodovar C, Lambrechts D, Mazzone M, Carmeliet P (2009) Role and therapeutic potential of VEGF in the nervous system. Physiol Rev 89:607–648

Saint-Geniez M, Maharaj AS, Walshe TE, Tucker BA, Sekiyama E, Kurihara T, Darland DC, Young MJ, D'Amore PA (2008) Endogenous VEGF is required for visual function: evidence for a survival role on Muller cells and photoreceptors. PLoS One 3:e3554

Schneider M, van Geyte K, Fraisl P, Kiss J, Aragones J, Mazzone M, Mairbaurl H, Debock K, Ho Jeoung N, Mollenhauer M, Georgiadou M, Bishop T, Roncal C, Sutherland A, Jordan B, Gallez B, Weitz J, Harris RA, Maxwell P, Baes M, Ratcliffe P, Carmeliet P (2009) Loss or silencing of the PHD1 prolyl hydroxylase protects livers of mice against ischemia/reperfusion injury. Gastroenterology 138:1143–1154

Schofield CJ, Ratcliffe PJ (2004) Oxygen sensing by HIF hydroxylases. Nat Rev Mol Cell Biol 5:343–354

Schubert D, Soucek T, Blouw B (2009) The induction of HIF-1 reduces astrocyte activation by amyloid beta peptide. Eur J Neurosci 29:1323–1334

Sebastia J, Kieran D, Breen B, King MA, Netteland DF, Joyce D, Fitzpatrick SF, Taylor CT, Prehn JH (2009) Angiogenin protects motoneurons against hypoxic injury. Cell Death Differ 16:1238–1247

Segura I, De Smet F, Hohensinner PJ, Almodovar CR, Carmeliet P (2009) The neurovascular link in health and disease: an update. Trends Mol Med 15:439–451

Selak MA, Armour SM, MacKenzie ED, Boulahbel H, Watson DG, Mansfield KD, Pan Y, Simon MC, Thompson CB, Gottlieb E (2005) Succinate links TCA cycle dysfunction to oncogenesis by inhibiting HIF-alpha prolyl hydroxylase. Cancer Cell 7:77–85

Semenza GL (2003) Targeting HIF-1 for cancer therapy. Nat Rev Cancer 3:721–732

Semenza GL (2007) Life with oxygen. Science 318:62–64

Semenza GL (2009) Regulation of cancer cell metabolism by hypoxia-inducible factor 1. Semin Cancer Biol 19:12–16

Semenza GL (2010a) Defining the role of hypoxia-inducible factor 1 in cancer biology and therapeutics. Oncogene 29(5):625–634

Semenza GL (2010b) HIF-1: upstream and downstream of cancer metabolism. Curr Opin Genet Dev 20(1):51–56

Shaked Y, Ciarrocchi A, Franco M, Lee CR, Man S, Cheung AM, Hicklin DJ, Chaplin D, Foster FS, Benezra R, Kerbel RS (2006) Therapy-induced acute recruitment of circulating endothelial progenitor cells to tumors. Science 313:1785–1787

Sharp FR, Bernaudin M (2004) HIF1 and oxygen sensing in the brain. Nat Rev Neurosci 5: 437–448

Sheehan JP, Swerdlow RH, Miller SW, Davis RE, Parks JK, Parker WD, Tuttle JB (1997) Calcium homeostasis and reactive oxygen species production in cells transformed by mitochondria from individuals with sporadic Alzheimer's disease. J Neurosci 17:4612–4622

Siddiq A, Ayoub IA, Chavez JC, Aminova L, Shah S, LaManna JC, Patton SM, Connor JR, Cherny RA, Volitakis I, Bush AI, Langsetmo I, Seeley T, Gunzler V, Ratan RR (2005) Hypoxia-inducible factor prolyl 4-hydroxylase inhibition. A target for neuroprotection in the central nervous system. J Biol Chem 280:41732–41743

Siddiq A, Aminova LR, Troy CM, Suh K, Messer Z, Semenza GL, Ratan RR (2009) Selective inhibition of hypoxia-inducible factor (HIF) prolyl-hydroxylase 1 mediates neuroprotection against normoxic oxidative death via HIF- and CREB-independent pathways. J Neurosci 29:8828–8838

Smith EE, Greenberg SM (2009) Beta-amyloid, blood vessels, and brain function. Stroke 40:2601–2606

Soucek T, Cumming R, Dargusch R, Maher P, Schubert D (2003) The regulation of glucose metabolism by HIF-1 mediates a neuroprotective response to amyloid beta peptide. Neuron 39:43–56

Stenzel-Poore MP, Stevens SL, Xiong Z, Lessov NS, Harrington CA, Mori M, Meller R, Rosenzweig HL, Tobar E, Shaw TE, Chu X, Simon RP (2003) Effect of ischaemic preconditioning on genomic response to cerebral ischaemia: similarity to neuroprotective strategies in hibernation and hypoxia-tolerant states. Lancet 362:1028–1037

Storkebaum E, Carmeliet P (2004) VEGF: a critical player in neurodegeneration. J Clin Invest 113:14–18

Storkebaum E, Lambrechts D, Dewerchin M, Moreno-Murciano MP, Appelmans S, Oh H, Van Damme P, Rutten B, Man WY, De Mol M, Wyns S, Manka D, Vermeulen K, Van Den Bosch L, Mertens N, Schmitz C, Robberecht W, Conway EM, Collen D, Moons L, Carmeliet P (2005) Treatment of motoneuron degeneration by intracerebroventricular delivery of VEGF in a rat model of ALS. Nat Neurosci 8:85–92

Sullivan R, Graham CH (2007) Hypoxia-driven selection of the metastatic phenotype. Cancer Metastasis Rev 26:319–331

Sullivan R, Graham CH (2009) Hypoxia prevents etoposide-induced DNA damage in cancer cells through a mechanism involving hypoxia-inducible factor 1. Mol Cancer Ther 8:1702–1713

Sun X, He G, Qing H, Zhou W, Dobie F, Cai F, Staufenbiel M, Huang LE, Song W (2006) Hypoxia facilitates Alzheimer's disease pathogenesis by up-regulating BACE1 gene expression. Proc Natl Acad Sci USA 103:18727–18732

Swietach P, Vaughan-Jones RD, Harris AL (2007) Regulation of tumor pH and the role of carbonic anhydrase 9. Cancer Metastasis Rev 26:299–310

Takeda K, Cowan A, Fong GH (2007) Essential role for prolyl hydroxylase domain protein 2 in oxygen homeostasis of the adult vascular system. Circulation 116:774–781

Tanzi RE, Moir RD, Wagner SL (2004) Clearance of Alzheimer's Abeta peptide: the many roads to perdition. Neuron 43:605–608

Tarkowski E, Issa R, Sjogren M, Wallin A, Blennow K, Tarkowski A, Kumar P (2002) Increased intrathecal levels of the angiogenic factors VEGF and TGF-beta in Alzheimer's disease and vascular dementia. Neurobiol Aging 23:237–243

Taylor CT, Pouyssegur J (2007) Oxygen, hypoxia, and stress. Ann N Y Acad Sci 1113:87–94

Tennant DA, Frezza C, MacKenzie ED, Nguyen QD, Zheng L, Selak MA, Roberts DL, Dive C, Watson DG, Aboagye EO, Gottlieb E (2009) Reactivating HIF prolyl hydroxylases under hypoxia results in metabolic catastrophe and cell death. Oncogene 28:4009–4021

Thomas T, Thomas G, McLendon C, Sutton T, Mullan M (1996) beta-Amyloid-mediated vasoactivity and vascular endothelial damage. Nature 380:168–171

Vander Heiden MG, Cantley LC, Thompson CB (2009) Understanding the Warburg effect: the metabolic requirements of cell proliferation. Science 324:1029–1033

Vasseur S, Afzal S, Tardivel-Lacombe J, Park DS, Iovanna JL, Mak TW (2009) DJ-1/PARK7 is an important mediator of hypoxia-induced cellular responses. Proc Natl Acad Sci USA 106: 1111–1116

Vaupel P, Mayer A (2007) Hypoxia in cancer: significance and impact on clinical outcome. Cancer Metastasis Rev 26:225–239

Wang R, Zhang YW, Zhang X, Liu R, Hong S, Xia K, Xia J, Zhang Z, Xu H (2006) Transcriptional regulation of APH-1A and increased gamma-secretase cleavage of APP and Notch by HIF-1 and hypoxia. FASEB J 20:1275–1277

Wang Y, Mao XO, Xie L, Banwait S, Marti HH, Greenberg DA, Jin K (2007) Vascular endothelial growth factor overexpression delays neurodegeneration and prolongs survival in amyotrophic lateral sclerosis mice. J Neurosci 27:304–307

Wouters BG, Koritzinsky M (2008) Hypoxia signalling through mTOR and the unfolded protein response in cancer. Nat Rev Cancer 8:851–864

Wouters BG, van den Beucken T, Magagnin MG, Lambin P, Koumenis C (2004) Targeting hypoxia tolerance in cancer. Drug Resist Updat 7:25–40

Xenaki G, Ontikatze T, Rajendran R, Stratford IJ, Dive C, Krstic-Demonacos M, Demonacos C (2008) PCAF is an HIF-1alpha cofactor that regulates p53 transcriptional activity in hypoxia. Oncogene 27:5785–5796

Xu L, Duda DG, di Tomaso E, Ancukiewicz M, Chung DC, Lauwers GY, Samuel R, Shellito P, Czito BG, Lin PC, Poleski M, Bentley R, Clark JW, Willett CG, Jain RK (2009) Direct evidence that bevacizumab, an anti-VEGF antibody, up-regulates SDF1alpha, CXCR4, CXCL6, and neuropilin 1 in tumors from patients with rectal cancer. Cancer Res 69:7905–7910

Xue J, Li X, Jiao S, Wei Y, Wu G, Fang J (2010) Prolyl hydroxylase-3 is down-regulated in colorectal cancer cells and inhibits IKKbeta independent of hydroxylase activity. Gastroenterology 138(2):606–615

Yan H, Parsons DW, Jin G, McLendon R, Rasheed BA, Yuan W, Kos I, Batinic-Haberle I, Jones S, Riggins GJ, Friedman H, Friedman A, Reardon D, Herndon J, Kinzler KW, Velculescu VE, Vogelstein B, Bigner DD (2009a) IDH1 and IDH2 mutations in gliomas. N Engl J Med 360:765–773

Yan M, Rayoo M, Takano EA, Thorne H, Fox SB (2009b) BRCA1 tumours correlate with a HIF-1 alpha phenotype and have a poor prognosis through modulation of hydroxylase enzyme profile expression. Br J Cancer 101:1168–1174

Yao J, Irwin RW, Zhao L, Nilsen J, Hamilton RT, Brinton RD (2009) Mitochondrial bioenergetic deficit precedes Alzheimer's pathology in female mouse model of Alzheimer's disease. Proc Natl Acad Sci USA 106:14670–14675

Yu JL, Rak JW, Coomber BL, Hicklin DJ, Kerbel RS (2002) Effect of p53 status on tumor response to antiangiogenic therapy. Science 295:1526–1528

Zacchigna S, Lambrechts D, Carmeliet P (2008) Neurovascular signalling defects in neurodegeneration. Nat Rev Neurosci 9:169–181

Zhang X, Zhou K, Wang R, Cui J, Lipton SA, Liao FF, Xu H, Zhang YW (2007) Hypoxia-inducible factor 1alpha (HIF-1alpha)-mediated hypoxia increases BACE1 expression and beta-amyloid generation. J Biol Chem 282:10873–10880

Zhang Q, Gu J, Li L, Liu J, Luo B, Cheung HW, Boehm JS, Ni M, Geisen C, Root DE, Polyak K, Brown M, Richardson AL, Hahn WC, Kaelin WG Jr, Bommi-Reddy A (2009) Control of cyclin D1 and breast tumorigenesis by the EglN2 prolyl hydroxylase. Cancer Cell 16:413–424

Zhao S, Lin Y, Xu W, Jiang W, Zha Z, Wang P, Yu W, Li Z, Gong L, Peng Y, Ding J, Lei Q, Guan KL, Xiong Y (2009) Glioma-derived mutations in IDH1 dominantly inhibit IDH1 catalytic activity and induce HIF-1alpha. Science 324:261–265

Zheng X, Linke S, Dias JM, Gradin K, Wallis TP, Hamilton BR, Gustafsson M, Ruas JL, Wilkins S, Bilton RL, Brismar K, Whitelaw ML, Pereira T, Gorman JJ, Ericson J, Peet DJ, Lendahl U, Poellinger L (2008) Interaction with factor inhibiting HIF-1 defines an additional mode of cross-coupling between the Notch and hypoxia signaling pathways. Proc Natl Acad Sci USA 105:3368–3373

Zhong Z, Deane R, Ali Z, Parisi M, Shapovalov Y, O'Banion MK, Stojanovic K, Sagare A, Boillee S, Cleveland DW, Zlokovic BV (2008) ALS-causing SOD1 mutants generate vascular changes prior to motor neuron degeneration. Nat Neurosci 11:420–422

Zhong Z, Ilieva H, Hallagan L, Bell R, Singh I, Paquette N, Thiyagarajan M, Deane R, Fernandez JA, Lane S, Zlokovic AB, Liu T, Griffin JH, Chow N, Castellino FJ, Stojanovic K, Cleveland DW, Zlokovic BV (2009) Activated protein C therapy slows ALS-like disease in mice by transcriptionally inhibiting SOD1 in motor neurons and microglia cells. J Clin Invest 119:3437–3449

Zhou D, Matchett GA, Jadhav V, Dach N, Zhang JH (2008) The effect of 2-methoxyestradiol, a HIF-1 alpha inhibitor, in global cerebral ischemia in rats. Neurol Res 30:268–271

Zlokovic BV (2005) Neurovascular mechanisms of Alzheimer's neurodegeneration. Trends Neurosci 28:202–208

Zlokovic BV (2008) The blood-brain barrier in health and chronic neurodegenerative disorders. Neuron 57:178–201

Zuniga RM, Torcuator R, Jain R, Anderson J, Doyle T, Ellika S, Schultz L, Mikkelsen T (2009) Efficacy, safety and patterns of response and recurrence in patients with recurrent high-grade gliomas treated with bevacizumab plus irinotecan. J Neurooncol 91:329–336

Hypoxia-Inducible Factors as Essential Regulators of Inflammation

Hongxia Z. Imtiyaz and M. Celeste Simon

Contents

Abstract Myeloid cells provide important functions in low oxygen (O_2) environments created by pathophysiological conditions, including sites of infection,

H.Z. Imtiyaz
Abramson Family Cancer Research Institute, University of Pennsylvania, 438 BRB II/III, 421 Curie Boulevard, Philadelphia, PA 19104-6160, USA
Howard Hughes Medical Institute, Chevy Chase, MD, USA

M.C. Simon (✉)
Abramson Family Cancer Research Institute, University of Pennsylvania, 438 BRB II/III, 421 Curie Boulevard, Philadelphia, PA 19104-6160, USA
Howard Hughes Medical Institute, Chevy Chase, MD, USA
e-mail: celeste2@mail.med.upenn.edu

Department of Cell and Developmental Biology, University of Pennsylvania School of Medicine, Philadelphia, PA 19104, USA

M. Celeste Simon (ed.), *Diverse Effects of Hypoxia on Tumor Progression*,
Current Topics in Microbiology and Immunology 345, DOI 10.1007/82_2010_74
© Springer-Verlag Berlin Heidelberg 2010, published online: 2 June 2010

inflammation, tissue injury, and solid tumors. Hypoxia-inducible factors (HIFs) are principle regulators of hypoxic adaptation, regulating gene expression involved in glycolysis, erythropoiesis, angiogenesis, proliferation, and stem cell function under low O_2. Interestingly, increasing evidence accumulated over recent years suggests an additional important regulatory role for HIFs in inflammation. In macrophages, HIFs not only regulate glycolytic energy generation, but also optimize innate immunity, control pro-inflammatory gene expression, mediate bacterial killing and influence cell migration. In neutrophils, HIF-1α promotes survival under O_2-deprived conditions and mediates blood vessel extravasation by modulating β_2 integrin expression. Additionally, HIFs contribute to inflammatory functions in various other components of innate immunity, such as dendritic cells, mast cells, and epithelial cells. This review will dissect the role of each HIF isoform in myeloid cell function and discuss their impact on acute and chronic inflammatory disorders. Currently, intensive studies are being conducted to illustrate the connection between inflammation and tumorigenesis. Detailed investigation revealing interaction between microenvironmental factors such as hypoxia and immune cells is needed. We will also discuss how hypoxia and HIFs control properties of tumor-associated macrophages and their relationship to tumor formation and progression.

1 Introduction

Tissue O_2 concentrations are typically maintained by homeostatic mechanisms operating at the cellular, organ, and systemic levels. *In vivo*, O_2 tension varies from 2.5 to 9% in most healthy tissues. However, inflamed or diseased tissues can be deprived of O_2 due to vascular damage, intensive metabolic activity of bacteria and other pathogens, and large numbers of infiltrating cells, leading to O_2 levels of less than 1% (Leek and Harris 2002; Lewis et al. 1999). As the front line of a body's defense, myeloid cells are required to function in this hypoxic microenvironment to combat infection, mediate inflammation, promote adaptive immunity and perform tissue repair functions (Lewis et al. 1999). These cells are unique in that they are well-adapted to hypoxia both metabolically and functionally. For example, neutrophils naturally operate under a pro-glycolytic program, and low O_2 endows neutrophils with a survival advantage over normoxic conditions (Hannah et al. 1995; Walmsley et al. 2005a, b). Macrophages specifically infiltrate hypoxic tissues, switch their metabolic program to glycolysis, resist apoptotic stimuli, and respond to O_2 deprivation by altering gene expression to maximize their biological properties (Cramer et al. 2003; Murdoch et al. 2004).

Cells adapt to hypoxia by shifting their energy generation pathway from aerobic oxidative phosphorylation to anaerobic glycolysis, and recovering blood supply via stimulation of erythrocyte production and the generation of new blood vessels (Semenza 2009). In addition to these established cellular responses, another aspect of hypoxia in connection with inflammation has been increasingly appreciated over recent years. A low O_2 microenvironment appears to actually amplify myeloid cell-mediated inflammatory responses, contributing to a highly inflamed state. During O_2 deprivation, many cellular responses are primarily regulated by HIFs. In pathological settings which involve inflammation or innate defense processes, HIFs are required to control programs associated with a broad range of myeloid cell functions (Cramer et al. 2003; Jantsch et al. 2008; Peyssonnaux et al. 2005, 2007; Walmsley et al. 2005b). HIFs are, therefore, essential regulators of inflammation and innate immunity, as will be detailed below. The relationship between inflammation and cancer, and how hypoxia and HIFs contribute to these processes will also be discussed. Adaptation of tumor-associated macrophages (TAMs) to hypoxic tumor microenvironments and their influences on tumor phenotypes will be highlighted.

2 Regulation of HIF Transcriptional Pathways

2.1 Oxygen-Dependent HIF Activity

In mammalian cells, hypoxic adaptation is primarily regulated by master transcriptional factors, called HIFs, whose activity is based on the post-translational modification and stability of their α subunits (HIF-1α and HIF-2α). In O_2 replete cells, prolyl hydroxylases (PHD-1, -2, and -3) modify the α subunit at two conserved prolines, resulting in polyubiquitylation via a specific von Hippel-Lindau (pVHL)-E3 ligase complex and subsequent degradation by proteasomes (Ivan et al. 2001; Jaakkola et al. 2001; Masson et al. 2001; Yu et al. 2001). Meanwhile, asparaginyl hydroxylation of HIF-α by factor inhibiting HIF (FIH) prevents its interaction with the co-activator p300/CBP, resulting in transcriptional inactivation under normoxia (Lando et al. 2002; Mahon et al. 2001; Sang et al. 2002). At sites of reduced O_2 tension, PHD and FIH hydroxylase activities are reduced. Stabilized α subunits then translocate to the nucleus, form dimers with constitutive HIF-1β (also known as the aryl hydrocarbon receptor nuclear translocator (ARNT)) and bind to co-activators, permitting transcriptional activation of many hypoxia-response element (HRE)-bearing genes encoding metabolic, angiogenic and metastatic factors (Covello and Simon 2004).

2.2 Inflammatory Stimuli-Induced HIF Activity

Besides O_2-dependent activation pathways, HIFs are also induced by inflammatory cytokines, growth factors and bacterial products at normoxic conditions, although the underlying molecular pathways are not fully revealed. Pro-inflammatory cytokines TNF-α and IL-1β have been shown to increase accumulation and transcriptional activity of HIF-1α. TNF-α-induced HIF-1α stimulation requires NF-κB at the level of HIF-1α protein stabilization without affecting its mRNA level (Jung et al. 2003a; Zhou et al. 2003). Similarly, IL-1β acts on HIF-1α protein stability by triggering NF-κB activity and inhibiting VHL-mediated protein degradation (Jung et al. 2003b). Moreover, TGF-β1 enhances HIF-1α protein stability by inhibiting PHD2 expression, via Smads (McMahon et al. 2006). The fact that HIF can be activated in response to inflammatory cytokines indicates HIF may play an important role in inflammation.

In addition to cytokines, bacteria and bacterial products such as lipopolysaccharide (LPS) also stimulate HIF-1α activity under normal O_2 levels. Several pathways have been reported to be involved in this process, including NF-κB (Fang et al. 2009; Frede et al. 2006; Nishi et al. 2008; Rius et al. 2008), ROS (Nishi et al. 2008), PHDs (Peyssonnaux et al. 2007) and p42/p44 mitogen-activated protein kinases (MAPKs) (Frede et al. 2006). The implication of NF-κB in this process has been controversial. Frede et al. reported that LPS induces HIF-1α mRNA expression in human monocytes through NF-κB binding to the promoter of the *HIF-1α* gene (Frede et al. 2006). Using IKK-β mutant macrophages, it was also shown that NF-κB is responsible for HIF-1α transcription and protein stability, and that *IKK-β* deficiency results in decreased expression of HIF targets such as glucose transporter-1 (Glut-1). In contrast, other studies demonstrated HIF-1α induction by LPS is not dependent on NF-κB activity (Fang et al. 2009; Nishi et al. 2008), but rather ROS generation (Nishi et al. 2008). Additionally, it has been shown that LPS increases HIF-1α protein accumulation through decreasing PHD2 and PHD3 levels in macrophages in a toll-like receptor-4 (TLR4)-dependent manner (Peyssonnaux et al. 2007). Future studies elucidating the crosstalk between HIFs and NF-κB are required given the importance of these two transcriptional factors in regulating hypoxic response and inflammation, respectively.

3 Role of HIFs in Myeloid Cell Functions

3.1 HIF Regulates Macrophage Activity

Macrophages display a variety of functions depending upon the type of stimulus presented in the local environment (Gordon 2003; Mosser and Edwards 2008). Interestingly, these cells accumulate in large numbers within O_2-deprived areas in various diseases such as bacterial infections, atherosclerosis, rheumatoid arthritis (RA), wounds and solid tumors, suggesting that hypoxic responses regulate macrophage biological activities (Murdoch et al. 2005). Exposure to hypoxia markedly changes macrophage gene expression profiles, resulting in the up-regulation of surface receptors (e.g., CXCR4 and Glut-1) and pro-angiogenic factors (Fang et al. 2009). Two α subunits, HIF-1α and HIF-2α, have been demonstrated to promote the expression of most O_2-regulated genes (Covello and Simon 2004). Whereas HIF-1α appears to be expressed ubiquitously, HIF-2α is expressed in a more tissue-restricted manner (Covello and Simon 2004). In macrophages, both HIF-1α and HIF-2α expression are induced in response to hypoxia *in vitro*

(Burke et al. 2002; Griffiths et al. 2000). Moreover, HIF-1α appears to be required for macrophage maturation (Fang et al. 2009; Oda et al. 2006). Interestingly, HIF-2α protein is readily detected *in vivo* in bone marrow macrophages and has been shown to be highly expressed in TAMs found in various human cancers (Talks et al. 2000). To elucidate the relative contribution of each HIF-α in the regulation of hypoxia-induced macrophage gene expression, siRNA-mediated knockdown of individual HIF-α subunits were performed in human monocyte-derived macrophages (Fang et al. 2009). Whereas HIF-1α and HIF-2α regulate expression of multiple common genes such as CXCR4, Glut-1, adrenomedulin (ADM) and STAT-4, expression of certain genes such as adenosine A2a (ADORA2A) and ICAM1 was only modulated by HIF-2α. Furthermore, over-expression of HIF-2α, but not HIF-1α, in normoxic human macrophages leads to enhanced transcription of pro-angiogenic genes including *VEGF*, *IL-8*, platelet-derived growth factor β (PDGFB) and angiopoietin-like 4 (*ANGPTL4*) (White et al. 2004). Collectively, these studies suggest HIF isoforms may play overlapping, but also distinct, roles in macrophage adaptation to low O_2.

To investigate macrophage biological properties, myeloid-specific ablation of the HIF-1α subunit in mice was created by crossing the floxed *Hif-1α* allele with a lysozyme M *cre* line (Cramer et al. 2003). This study demonstrated a dominant role for HIF-1α in regulating glycolysis in macrophages (Cramer et al. 2003) as HIF-1α deficiency results in a dramatically reduced ATP pool. This is consistent with other studies demonstrating that HIF-1α exclusively controls glycolysis (Hu et al. 2003). The metabolic defect in HIF-1α deletion in macrophages results in impairment of energy-demanding processes such as aggregation, migration and invasion (Cramer et al. 2003).

In addition to its key role in regulating metabolism and energy generation, fundamental work by Cramer et al. showed that HIF-1α mediates macrophage inflammatory responses. Compared to control mice, myeloid HIF-1α-null mice displayed reduced acute skin inflammation triggered by 12-O-tetradecanoylphorbol-13-acetate (TPA), as evidenced by decreased edema and leukocyte infiltration (Cramer et al. 2003). When induced to develop arthritis, these mice also showed compromised synovial infiltration, pannus formation and cartilage destruction, suggesting ameliorated chronic inflammatory responses mediated by HIF-1α-deficient macrophages.

Many of the pro-inflammatory cytokine/chemokine genes are activated by hypoxic treatment in human primary macrophages. Compromised expression of IL-1β, CXCL8 and VEGF was observed in cells exhibiting reduced expression of either HIF-1α or HIF-2α, indicating both HIFs are important for macrophage cytokine expression (Fang et al. 2009). Given these cytokines/chemokines are also known to be NF-κB targets, the role of NF-κB in inducing their expression under low O_2 concentration has been evaluated. However, inactivation of NF-κB, either chemically or genetically, did not influence hypoxia-induced cytokine expression (Fang et al. 2009). This result suggests that HIFs, but not NF-κB, are important transcriptional effectors regulating the hypoxic gene expression of macrophages.

Innate immunity was also assessed in myeloid HIF-1α null mice by Peyssonnaux et al. (2005). The authors demonstrated that loss of myeloid HIF-1α resulted in decreased bacterial killing of group A *Streptococcus* and *P. aeruginosa* by macrophages *in vitro* and *in vivo* (Peyssonnaux et al. 2005), revealing the importance of myeloid HIF-1α in this process. Furthermore, exposure to these pathogens and LPS under normoxia induces HIF-1α activity in macrophages in a TLR-4 dependent fashion (Peyssonnaux et al. 2007). Finally, HIF-1α directly binds to the promoter of the *TLR4* locus and up-regulates *TLR4* expression during O_2 deprivation (Kim et al. 2009). Thus, the interdependent relationship between HIF-1α and TLR4 activation may result in a positive feedback loop, amplifying HIF responses under hypoxia and infection. Collectively, these findings suggest hypoxic stress at sites of inflammation both enhances sensitivity to infection by strengthening TLR4 signaling and promotes the defense capacity of macrophages by increasing HIF-1α levels (Fig. 1).

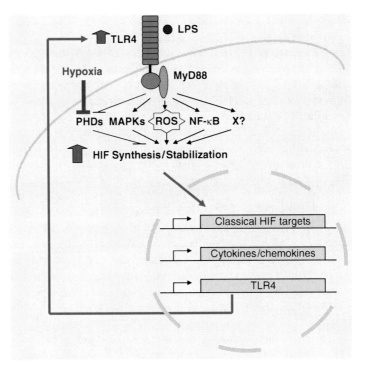

Fig. 1 Possible mechanisms for maximal inflammatory responses elicited by lipopolysaccharide (LPS) during hypoxia. In macrophages, LPS induces hypoxia-inducible factor (HIF) protein synthesis and stabilization in a TLR4/MyD88-dependent fashion. The signaling pathways for LPS-induced HIF stabilization possibly involve ROS, inhibition of prolyl hydroxylases (PHDs), mitogen-activated protein kinase (MAPK), NF-κB and/or unknown factors (X). Upon HIF protein accumulation, it translocates to the nucleus and stimulates expression of classical HIF target genes, inflammatory cytokines/chemokines, and TLR4. TLR4 further enhances HIF expression and transcriptional activity, resulting in a positive feedback loop. Together with the LPS-TLR4 pathway, hypoxia itself stabilizes HIF protein via inhibiting PHDs, serving to amplify HIF-mediated inflammatory responses during infection and O_2 deprivation

Given the importance of HIF-2α in macrophages implicated by gene expression studies (Fang et al. 2009; Griffiths et al. 2000; Talks et al. 2000; White et al. 2004), genetic experiments utilizing mouse models have now been performed to investigate the role of HIF-2α in various aspects of macrophage functions (Imtiyaz et al., manuscript submitted). A clear division of labor exists between the two HIF-α isoforms in these cells where HIF-2α is clearly important for inflammatory cytokine expression, macrophage migration, and responses to inflammatory stimuli.

3.2 HIF Function in Other Myeloid Cells

Neutrophils are important phagocytes which clear invading pathogens and mediate acute inflammation (Bredetean et al. 2007). As they rely on glycolysis to generate ATP, these cells seem to be well-suited to function in the hypoxic microenvironment naturally present in inflammatory lesions (Walmsley et al. 2005a). Studies of HIF-1α-deficient neutrophils revealed that neutrophils require HIF-1α to perform glycolysis (Cramer et al. 2003). Using murine bone marrow-derived neutrophils and human peripheral blood neutrophils, Walmsley et al. demonstrated that cells deficient in HIF-1α failed to resist apoptosis under hypoxic conditions (Walmsley et al. 2005b). Moreover, this HIF-mediated survival effect is also dependent on the activity of NF-κB, and is eliminated with treatment of NF-κB inhibitors (gliotoxin and parthenolide). Migration of neutrophils from circulation to sites of infection or tissue damage involves a process of selectin-mediated rolling and β_2 integrin-mediated adhesion to endothelium (Carlos and Harlan 1994). It has been demonstrated that HIF-1α regulates β_2 integrin (CD18 specifically) expression in these cells and thereby promotes neutrophil extravasation (Kong et al. 2004). HIF-2α is unlikely to play a role in neutrophil function as it is not expressed in this lineage (Walmsley et al. 2005b) (Imtiyaz et al., submitted).

Progress has been made in deciphering the role of HIF-1α in two other types of myeloid cells. As professional antigen presentation cells, dendritic cells (DCs) play a key role in linking innate and adaptive immunity. Recent work by Jantsch et al. has revealed that hypoxia and HIF-1α modulate DC maturation, activation and antigen-presenting functions (Jantsch et al. 2008). Although hypoxia alone did not activate DCs, hypoxia combined with LPS led to marked increases in expression of co-stimulatory molecules, pro-inflammatory cytokine synthesis and induction of lymphocyte proliferation compared with LPS alone. This DC activation was accompanied by HIF-1α protein accumulation and enhanced glycolytic activity. Moreover, knockdown of HIF-1α significantly reduced glucose uptake, inhibited maturation and led to an impaired capacity to stimulate allogeneic T cells (Jantsch et al. 2008). Mast cells are granulocytes implicated in allergy, and evidence regarding their roles in both innate and adaptive immunity is now emerging. Activation of HIF-1α in mast cells stimulates expression of VEGF, CXCL8, IL-6, and TNF (Jeong et al. 2003; Lee et al. 2008). It will be interesting to determine the expression and possible function of HIF-2α in both DCs and mast cells in future experiments.

4 Hypoxia, HIFs and Inflammatory Diseases

4.1 Sepsis

Sepsis is an aberrant host inflammatory response provoked by overwhelming infection or LPS. It leads to the potentially lethal systemic inflammatory response syndrome (SIRS) characterized by acute inflammation, hemodynamic compromise, multi-organ failure and even shock (Jean-Baptiste 2007; Parrillo 1993). Currently, sepsis is still the leading cause of mortality in intensive care units. The systemic effects of LPS are largely mediated by macrophages which produce a wide array of inflammatory cytokines (Jean-Baptiste 2007; Ulloa and Tracey 2005). Pro-inflammatory cytokines IL-1β, IL-12, TNF-α, and INF-γ have all been implicated in the toxic effects of endotoxemia, as neutralization of individual cytokines by specific antibodies protects mice from LPS-induced lethality (Dinarello 1991; Doherty et al. 1992; Heinzel 1990; Kumar et al. 1996; Tracey et al. 1987; Zisman et al. 1997). In contrast, the anti-inflammatory cytokine IL-10 has been proven to be beneficial (Howard et al. 1993; Nicoletti et al. 1997; Standiford et al. 1995). Studies of LPS-induced responses in myeloid HIF-targeted mice revealed that HIF-1α is important for the sepsis phenotype. HIF-1α deletion in myeloid cells led to reduced pro-inflammatory cytokines such as TNF-α, IL-1, and IL-12 (Peyssonnaux et al. 2007). In addition, HIF-1α contributes to the lethal effects of LPS as mice survived much longer when myeloid HIF-1α is absent. Moreover, HIF-1α deletion blocked LPS-induced hypotension and hypothermia caused by sepsis (Peyssonnaux et al. 2007). Whether HIF-2α ablation in macrophages results in a similar septic phenotype requires future study.

4.2 Atherosclerosis

Atherosclerosis is a chronic inflammatory response in the walls of arteries, in large part due to the accumulation of macrophages (known as foam cells) that take up oxidized low-density lipoproteins (Portugal et al. 2009). The build-up of "fatty streaks" by these fat-containing macrophages forms atherosclerotic plaques, leading to arterial stenosis and impaired perfusion (Zemplenyi et al. 1989). Using the hypoxia marker 7-(4'-(2-nitroimidazol-1-yl)-butyl)-theophylline, a previous study has shown zones of hypoxia are present in the plaques, probably due to impaired O_2 diffusion into these lesions (Bjornheden et al. 1999). At least two macrophage-derived products have been implicated in the plaque development. The first one is very low-density lipoprotein receptor (VLDLR) which has been shown to be expressed by macrophages in atherosclerotic lesions (Nakazato et al. 1996). In vitro, VLDLR levels are increased by hypoxia in macrophages (Nakazato et al. 2001), although whether this is dependent on HIF expression is currently unclear. The second factor is CXCL-8 (i.e., IL-8), a potent chemoattractant for

T lymphocytes and smooth muscle cells. Significant elevation in CXCL-8 production has been found in foam cells isolated from human atherosclerotic plaques compared with macrophages in culture (Liu et al. 1997). Of note, foam cells found in hypoxic zones displayed enhanced CXCL-8 levels in both rabbit and human atherosclerotic sites compared to the normal arterial wall (Rydberg et al. 2003). Moreover, hypoxia induces CXCL-8 expression in primary human macrophages, mediated by both HIF-1α and HIF-2α (Fang et al. 2009). Interestingly, expression of CXCR2, the receptor for CXCL-8, in macrophages significantly contributes to the progression of advanced atherosclerosis in mice (Boisvert et al. 2000), underscoring the importance of the CXCL8-CXCR2 signaling axis in this disease. It will be interesting to determine the genetic requirements for either HIF-1α or HIF-2α expression in macrophages using appropriate animal models of atherosclerosis.

4.3 Rheumatoid Arthritis

RA is another type of chronic inflammatory disorder that primarily attacks the joints, producing a synovitis that often progresses to destruction of bone and cartilage (Muz et al. 2009). Although the cause of RA is unknown, hypoxia has been suggested to contribute to its pathology (Muz et al. 2009; Sivakumar et al. 2008). Using microelectrodes and the hypoxia marker pimonidazole (PIMO) staining, reduced O_2 tension has been detected in the synovium of RA patients and animals (Peters et al. 2004; Sivakumar et al. 2008). The presence of hypoxia in RA joints is probably attributable to continuous synovial expansion which outstrips the blood-borne O_2 supply. In RA synovial membrane cultures which contain macrophages, lymphocytes and fibroblasts, hypoxia appears to be a potent stimulus for *VEGF* induction, a classical hypoxia-responsive gene. Moreover, macrophages in RA joints express factors such as VEGF, IL-1, TNF-α, CXCL-8, CXCL-12, Cox-2, and MMP-1 (Muz et al. 2009), some of which are known hypoxia-regulated factors. The precise mechanism for how hypoxia regulates these molecules in RA joints is unclear and requires further investigation. Interestingly, several of these factors (e.g., VEGF, IL-1, TNF-α, and CXCL-8) are known to promote angiogenesis, a characteristic of RA progression (Szekanecz et al. 1998). However, there appears to be a paradox in that abundant synovial vasculature is nonetheless associated with regions of synovial hypoxia. This may suggest that over-activation of the angiogenesis cascade by hypoxia results in formation of chaotic vessels with decreased blood flow, similar to that seen in many solid tumors.

As described above, in a mouse model of induced arthritis, it has been shown that myeloid HIF-1α activity is important for disease development (Cramer et al. 2003). To understand precisely how HIF-1α is involved in RA pathogenesis, it would be useful to isolate synovial macrophages to analyze HIF-dependent gene expression. It remains to be determined whether HIF-2α expression in myeloid cells (mainly macrophages) is also required in RA pathogenesis.

5 HIF Activities in Tumor-Associated Macrophages

5.1 Connecting Inflammation and Cancer

Since Rudolf Virchow's observation in 1863 that leukocytes infiltrate malignant tissues, suggesting cancers arise at sites of chronic inflammation, a relationship between inflammation and cancer has emerged. Epidemiological studies clearly demonstrate that ~15% of human cancer deaths are associated with chronic viral or bacterial infections. For example, human papillomaviruses, hepatitis B virus (HBV) and hepatitis C virus (HCV), and the bacterium *Helicobacter pylori* cause cervical cancer, hepatocellular carcinoma and gastric cancer, respectively (Mantovani et al. 2008). This effect is attributed to inflammatory cells and cytokines thought to establish an inflammatory microenvironment in tumors (Balkwill and Mantovani 2001). Interestingly, within tumors, macrophages represent a major component of infiltrating leukocytes (also including DCs, neutrophils, mast cells, and T cells) (Kelly et al. 1988; Leek et al. 1994), as well as the nontumor stromal cell compartment.

Clinically, increased TAM density correlates with poor patient prognosis (Pollard 2004; Bingle et al. 2002). Such correlative data are particularly convincing for breast (Leek et al. 1996), prostate (Lissbrant et al. 2000), cervical (Fujimoto et al. 2000) and ovarian cancers (Pollard 2004). Using mouse models of macrophage colony stimulating factor (M-CSF) mutations, Lin et al. demonstrated that macrophage-deficient animals showed marked decreases in the rate of tumor metastasis, although primary tumor growth rate was normal (Lin et al. 2001). The authors concluded that TAM abundance potentiates tumor progression. Furthermore, using hepatocyte-specific NF-κB inactivation models, several groups have indicated that pro-inflammatory cytokines produced by Kupffer cells (namely IL-6, TNF-α, and IL-1β) promote compensatory proliferation of hepatocytes, resulting in a significant increase in hepatocarcinogenesis (Luedde et al. 2007; Maeda et al. 2005; Naugler et al. 2007). Therefore, rather than limiting tumors, inflammation can actually promote tumor initiation, growth and metastasis.

5.2 Hypoxia and TAMs

Solid tumors contain large areas of hypoxia, exhibiting O_2 tensions between 0.1 and 1%. The presence of increased hypoxic domains correlates with poor prognosis (Vaupel et al. 2001), due to the relative resistance of hypoxic cells to conventional cancer therapies (Hockel and Vaupel 2001). Also, low O_2 promotes rapid angiogenesis and exerts pressure for the selection of mutant tumor cells with survival or growth advantages (Brown and Giaccia 1998). Interestingly, hypoxia and TAMs co-localize in tumor avascular or perinecrotic regions, indicating that TAMs specifically accumulate in O_2-deprived regions within tumors. To accomplish this, tumor cells produce chemokines CCL2 and CCL5, and the cytokine M-CSF which serve to recruit monocytes from the local vasculature to tumors. Upon tumor infiltration, monocytes differentiate into TAMs and migrate along the chemoattractant gradient generated by hypoxia (Murdoch et al. 2004). Increased expression of macrophage chemoattractants such as VEGF, endothelins, IL-8 and endothelial monocyte activating polypeptide II (EMAP II) occurs in hypoxic tumor cells. Thereafter, due to down-regulation of adhesion markers and chemo-attractant receptors, abrogation of chemotactic signal transduction, and the migra-tion inhibitory actions of MIF, TAMs decrease their motility and are subsequently immobilized in these O_2-deprived areas (Murdoch et al. 2004; Grimshaw and Balkwill 2001).

Studies of breast cancer have revealed a positive correlation between numbers of TAMs in hypoxic sites and levels of angiogenesis, lymph node involvement and poor prognosis (Grimshaw and Balkwill 2001). This suggests that O_2 depletion promotes TAM responses leading to the development of pro-tumor phenotypes. Under hypoxia, TAMs up-regulate hypoxia-inducible transcription factors, and activate expression programs that appear to be pro-angiogenenic, pro-tumor growth, pro-metastatic and immunosuppressive (Lewis and Pollard 2006; Pollard 2004; Sica et al. 2006).

5.3   HIF-2α Activity in TAMs

Although both HIF-1α and HIF-2α could be stabilized in hypoxic TAMs, work from Talks et al. showed that HIF-2α, in particular, is strongly expressed in these cells across a wide range of human tumors (Talks et al. 2000). To elucidate the impact of high TAM HIF-2α expression on tumor phenotypes and prognosis, clinical studies have been performed on human breast cancer by Leek et al. (2002). This investiga-tion revealed a positive correlation between the numbers of HIF-2α-expressing TAMs and poor prognosis. Moreover, high TAM HIF-2α levels are associated with increased tumor grade and tumor vascularity, suggesting HIF-2α expression in TAMs promotes tumor progression by improving angiogenesis (Leek et al. 2002). In another recent study, Kawanaka et al. investigated the significance of TAM HIF-2α expression in predicting survival and relapse on uterine cervical cancer patients undergoing radiotherapy (Kawanaka et al. 2008). Their results showed that increased numbers of HIF-2α-expressing TAMs are associated with poor disease-free survival and higher rate of local recurrence (Kawanaka et al. 2008). Given the clinical implications, studies determining the role of HIF-2α in TAMs using inflammation-associated tumor models and conditional knockout mouse lines is under investigation (Imtiyaz et al., manuscript submitted). It would be interesting to see how TAM HIF-2α affects tumor initiation, promotion and progression. Finally, determining whether HIF-1α is involved in TAM activities and tumorigenesis is certainly warranted.

6 Conclusions

Inflammation is a complex innate immune response elicited at sites experiencing infection, toxin exposure and injury. While proper inflammation helps to destroy infectious agents and restores tissue integrity, improper responses are harmful, leading to tissue destruction, vascular damage, and even organ failure in the case of sepsis. The connection between hypoxia and inflammation has become evident over the last decade, centering on the activity of HIFs. As ancient low-O_2 adaptation regulators expressed in all metazoan species, HIFs also confer responses to immune stresses. This is manifested by their ability to regulate cytokine expression, myeloid cell migration and effector functions. HIFs also act to further amplify these responses under hypoxia. Dysregulation of HIFs has been shown to result in various diseases, as revealed above. HIFs therefore represent both drug candidates and targets dependent on disease types. In an immunodeficient scenario, boosting HIF activity is expected to improve inflammation and effector functions to defeat infection. This could be achieved by inhibiting the activities of PHDs and pVHL protein which negatively regulate HIFs. Given that many inflammatory disorders cause either prolonged or exaggerated inflammatory responses, targeting HIF activity or its downstream genes would be an attractive strategy for therapeutical intervention. Moreover, inhibition of VEGF leading to vessel normalization, and thus tissue re-oxygenation, may also be helpful to treat chronic inflammation.

References

Balkwill F, Mantovani A (2001) Inflammation and cancer: back to Virchow? Lancet 357:539–545
Bingle L, Brown NJ, Lewis CE (2002) The role of tumour-associated macrophages in tumour progression: implications for new anticancer therapies. J Pathol 196:254–265
Bjornheden T, Levin M, Evaldsson M, Wiklund O (1999) Evidence of hypoxic areas within the arterial wall *in vivo*. Arterioscler Thromb Vasc Biol 19:870–876
Boisvert WA, Curtiss LK, Terkeltaub RA (2000) Interleukin-8 and its receptor CXCR2 in atherosclerosis. Immunol Res 21:129–137
Bredetean O, Ciochina AD, Mungiu OC (2007) The neutrophil in human pathology. Rev Med Chir Soc Med Nat Iasi 111:446–453
Brown JM, Giaccia AJ (1998) The unique physiology of solid tumors: opportunities (and problems) for cancer therapy. Cancer Res 58:1408–1416
Burke B, Tang N, Corke KP, Tazzyman D, Ameri K, Wells M, Lewis CE (2002) Expression of HIF-1alpha by human macrophages: implications for the use of macrophages in hypoxia-regulated cancer gene therapy. J Pathol 196:204–212
Carlos TM, Harlan JM (1994) Leukocyte-endothelial adhesion molecules. Blood 84:2068–2101
Covello KL, Simon MC (2004) HIFs, hypoxia, and vascular development. Curr Top Dev Biol 62:37–54
Cramer T, Yamanishi Y, Clausen BE, Forster I, Pawlinski R, Mackman N, Haase VH, Jaenisch R, Corr M, Nizet V et al (2003) HIF-1alpha is essential for myeloid cell-mediated inflammation. Cell 112:645–657
Dinarello CA (1991) The proinflammatory cytokines interleukin-1 and tumor necrosis factor and treatment of the septic shock syndrome. J Infect Dis 163:1177–1184

Doherty GM, Lange JR, Langstein HN, Alexander HR, Buresh CM, Norton JA (1992) Evidence for IFN-gamma as a mediator of the lethality of endotoxin and tumor necrosis factor-alpha. J Immunol 149:1666–1670

Fang HY, Hughes R, Murdoch C, Coffelt SB, Biswas SK, Harris AL, Johnson RS, Imityaz HZ, Simon MC, Fredlund E et al (2009) Hypoxia-inducible factors 1 and 2 are important transcriptional effectors in primary macrophages experiencing hypoxia. Blood 114:844–859

Frede S, Stockmann C, Freitag P, Fandrey J (2006) Bacterial lipopolysaccharide induces HIF-1 activation in human monocytes via p44/42 MAPK and NF-kappaB. Biochem J 396:517–527

Fujimoto J, Sakaguchi H, Aoki I, Tamaya T (2000) Clinical implications of expression of interleukin 8 related to angiogenesis in uterine cervical cancers. Cancer Res 60:2632–2635

Gordon S (2003) Alternative activation of macrophages. Nat Rev Immunol 3:23–35

Griffiths L, Binley K, Iqball S, Kan O, Maxwell P, Ratcliffe P, Lewis C, Harris A, Kingsman S, Naylor S (2000) The macrophage – a novel system to deliver gene therapy to pathological hypoxia. Gene Ther 7:255–262

Grimshaw MJ, Balkwill FR (2001) Inhibition of monocyte and macrophage chemotaxis by hypoxia and inflammation – a potential mechanism. Eur J Immunol 31:480–489

Hannah S, Mecklenburgh K, Rahman I, Bellingan GJ, Greening A, Haslett C, Chilvers ER (1995) Hypoxia prolongs neutrophil survival *in vitro*. FEBS Lett 372:233–237

Heinzel FP (1990) The role of IFN-gamma in the pathology of experimental endotoxemia. J Immunol 145:2920–2924

Hockel M, Vaupel P (2001) Tumor hypoxia: definitions and current clinical, biologic, and molecular aspects. J Natl Cancer Inst 93:266–276

Howard M, Muchamuel T, Andrade S, Menon S (1993) Interleukin 10 protects mice from lethal endotoxemia. J Exp Med 177:1205–1208

Hu CJ, Wang LY, Chodosh LA, Keith B, Simon MC (2003) Differential roles of hypoxia-inducible factor 1alpha (HIF-1alpha) and HIF-2alpha in hypoxic gene regulation. Mol Cell Biol 23:9361–9374

Ivan M, Kondo K, Yang H, Kim W, Valiando J, Ohh M, Salic A, Asara JM, Lane WS, Kaelin WG Jr (2001) HIFalpha targeted for VHL-mediated destruction by proline hydroxylation: implications for O_2 sensing. Science 292:464–468

Jaakkola P, Mole DR, Tian YM, Wilson MI, Gielbert J, Gaskell SJ, Kriegsheim A, Hebestreit HF, Mukherji M, Schofield CJ et al (2001) Targeting of HIF-alpha to the von Hippel-Lindau ubiquitylation complex by O2-regulated prolyl hydroxylation. Science 292:468–472

Jantsch J, Chakravortty D, Turza N, Prechtel AT, Buchholz B, Gerlach RG, Volke M, Glasner J, Warnecke C, Wiesener MS et al (2008) Hypoxia and hypoxia-inducible factor-1 alpha modulate lipopolysaccharide-induced dendritic cell activation and function. J Immunol 180:4697–4705

Jean-Baptiste E (2007) Cellular mechanisms in sepsis. J Intensive Care Med 22:63–72

Jeong HJ, Chung HS, Lee BR, Kim SJ, Yoo SJ, Hong SH, Kim HM (2003) Expression of proinflammatory cytokines via HIF-1alpha and NF-kappaB activation on desferrioxamine-stimulated HMC-1 cells. Biochem Biophys Res Commun 306:805–811

Jung Y, Isaacs JS, Lee S, Trepel J, Liu ZG, Neckers L (2003a) Hypoxia-inducible factor induction by tumour necrosis factor in normoxic cells requires receptor-interacting protein-dependent nuclear factor kappa B activation. Biochem J 370:1011–1017

Jung YJ, Isaacs JS, Lee S, Trepel J, Neckers L (2003b) IL-1beta-mediated up-regulation of HIF-1alpha via an NFkappaB/COX-2 pathway identifies HIF-1 as a critical link between inflammation and oncogenesis. Faseb J 17:2115–2117

Kawanaka T, Kubo A, Ikushima H, Sano T, Takegawa Y, Nishitani H (2008) Prognostic significance of HIF-2alpha expression on tumor infiltrating macrophages in patients with uterine cervical cancer undergoing radiotherapy. J Med Invest 55:78–86

Kelly PM, Davison RS, Bliss E, McGee JO (1988) Macrophages in human breast disease: a quantitative immunohistochemical study. Br J Cancer 57:174–177

Kim SY, Choi YJ, Joung SM, Lee BH, Jung YS, Lee JY (2009) Hypoxic stress up-regulates the expression of toll-like receptor 4 in macrophages via hypoxia-inducible factor. Immunology 4:516–524

Kong T, Eltzschig HK, Karhausen J, Colgan SP, Shelley CS (2004) Leukocyte adhesion during hypoxia is mediated by HIF-1-dependent induction of beta2 integrin gene expression. Proc Natl Acad Sci USA 101:10440–10445

Kumar A, Thota V, Dee L, Olson J, Uretz E, Parrillo JE (1996) Tumor necrosis factor alpha and interleukin 1beta are responsible for *in vitro* myocardial cell depression induced by human septic shock serum. J Exp Med 183:949–958

Lando D, Peet DJ, Whelan DA, Gorman JJ, Whitelaw ML (2002) Asparagine hydroxylation of the HIF transactivation domain a hypoxic switch. Science 295:858–861

Lee KS, Kim SR, Park SJ, Min KH, Lee KY, Choe YH, Park SY, Chai OH, Zhang X, Song CH, Lee YC (2008) Mast cells can mediate vascular permeability through regulation of the PI3K-HIF-1alpha-VEGF axis. Am J Respir Crit Care Med 178:787–797

Leek RD, Harris AL (2002) Tumor-associated macrophages in breast cancer. J Mammary Gland Biol Neoplasia 7:177–189

Leek RD, Harris AL, Lewis CE (1994) Cytokine networks in solid human tumors: regulation of angiogenesis. J Leukoc Biol 56:423–435

Leek RD, Lewis CE, Whitehouse R, Greenall M, Clarke J, Harris AL (1996) Association of macrophage infiltration with angiogenesis and prognosis in invasive breast carcinoma. Cancer Res 56:4625–4629

Leek RD, Talks KL, Pezzella F, Turley H, Campo L, Brown NS, Bicknell R, Taylor M, Gatter KC, Harris AL (2002) Relation of hypoxia-inducible factor-2 alpha (HIF-2 alpha) expression in tumor-infiltrative macrophages to tumor angiogenesis and the oxidative thymidine phosphorylase pathway in Human breast cancer. Cancer Res 62:1326–1329

Lewis CE, Pollard JW (2006) Distinct role of macrophages in different tumor microenvironments. Cancer Res 66:605–612

Lewis JS, Lee JA, Underwood JC, Harris AL, Lewis CE (1999) Macrophage responses to hypoxia: relevance to disease mechanisms. J Leukoc Biol 66:889–900

Lin EY, Nguyen AV, Russell RG, Pollard JW (2001) Colony-stimulating factor 1 promotes progression of mammary tumors to malignancy. J Exp Med 193:727–740

Lissbrant IF, Stattin P, Wikstrom P, Damber JE, Egevad L, Bergh A (2000) Tumor associated macrophages in human prostate cancer: relation to clinicopathological variables and survival. Int J Oncol 17:445–451

Liu Y, Hulten LM, Wiklund O (1997) Macrophages isolated from human atherosclerotic plaques produce IL-8, and oxysterols may have a regulatory function for IL-8 production. Arterioscler Thromb Vasc Biol 17:317–323

Luedde T, Beraza N, Kotsikoris V, van Loo G, Nenci A, De Vos R, Roskams T, Trautwein C, Pasparakis M (2007) Deletion of NEMO/IKKgamma in liver parenchymal cells causes steatohepatitis and hepatocellular carcinoma. Cancer Cell 11:119–132

Maeda S, Kamata H, Luo JL, Leffert H, Karin M (2005) IKKbeta couples hepatocyte death to cytokine-driven compensatory proliferation that promotes chemical hepatocarcinogenesis. Cell 121:977–990

Mahon PC, Hirota K, Semenza GL (2001) FIH-1: a novel protein that interacts with HIF-1alpha and VHL to mediate repression of HIF-1 transcriptional activity. Genes Dev 15:2675–2686

Mantovani A, Allavena P, Sica A, Balkwill F (2008) Cancer-related inflammation. Nature 454:436–444

Masson N, Willam C, Maxwell PH, Pugh CW, Ratcliffe PJ (2001) Independent function of two destruction domains in hypoxia-inducible factor-alpha chains activated by prolyl hydroxylation. EMBO J 20:5197–5206

McMahon S, Charbonneau M, Grandmont S, Richard DE, Dubois CM (2006) Transforming growth factor beta1 induces hypoxia-inducible factor-1 stabilization through selective inhibition of PHD2 expression. J Biol Chem 281:24171–24181

Mosser DM, Edwards JP (2008) Exploring the full spectrum of macrophage activation. Nat Rev Immunol 8:958–969

Murdoch C, Giannoudis A, Lewis CE (2004) Mechanisms regulating the recruitment of macrophages into hypoxic areas of tumors and other ischemic tissues. Blood 104:2224–2234

Murdoch C, Muthana M, Lewis CE (2005) Hypoxia regulates macrophage functions in inflammation. J Immunol 175:6257–6263

Muz B, Khan MN, Kiriakidis S, Paleolog EM (2009) Hypoxia. The role of hypoxia and HIF-dependent signalling events in rheumatoid arthritis. Arthritis Res Ther 11:201

Nakazato K, Ishibashi T, Shindo J, Shiomi M, Maruyama Y (1996) Expression of very low density lipoprotein receptor mRNA in rabbit atherosclerotic lesions. Am J Pathol 149: 1831–1838

Nakazato K, Ishibashi T, Nagata K, Seino Y, Wada Y, Sakamoto T, Matsuoka R, Teramoto T, Sekimata M, Homma Y, Maruyama Y (2001) Expression of very low density lipoprotein receptor mRNA in circulating human monocytes: its up-regulation by hypoxia. Atherosclerosis 155:439–444

Naugler WE, Sakurai T, Kim S, Maeda S, Kim K, Elsharkawy AM, Karin M (2007) Gender disparity in liver cancer due to sex differences in MyD88-dependent IL-6 production. Science 317:121–124

Nicoletti F, Mancuso G, Ciliberti FA, Beninati C, Carbone M, Franco S, Cusumano V (1997) Endotoxin-induced lethality in neonatal mice is counteracted by interleukin-10 (IL-10) and exacerbated by anti-IL-10. Clin Diagn Lab Immunol 4:607–610

Nishi K, Oda T, Takabuchi S, Oda S, Fukuda K, Adachi T, Semenza GL, Shingu K, Hirota K (2008) LPS induces hypoxia-inducible factor 1 activation in macrophage-differentiated cells in a reactive oxygen species-dependent manner. Antioxid Redox Signal 10:983–995

Oda T, Hirota K, Nishi K, Takabuchi S, Oda S, Yamada H, Arai T, Fukuda K, Kita T, Adachi T et al (2006) Activation of hypoxia-inducible factor 1 during macrophage differentiation. Am J Physiol Cell Physiol 291:C104–C113

Parrillo JE (1993) Pathogenetic mechanisms of septic shock. N Engl J Med 328:1471–1477

Peters CL, Morris CJ, Mapp PI, Blake DR, Lewis CE, Winrow VR (2004) The transcription factors hypoxia-inducible factor 1alpha and Ets-1 colocalize in the hypoxic synovium of inflamed joints in adjuvant-induced arthritis. Arthritis Rheum 50:291–296

Peyssonnaux C, Datta V, Cramer T, Doedens A, Theodorakis EA, Gallo RL, Hurtado-Ziola N, Nizet V, Johnson RS (2005) HIF-1alpha expression regulates the bactericidal capacity of phagocytes. J Clin Invest 115:1806–1815

Peyssonnaux C, Cejudo-Martin P, Doedens A, Zinkernagel AS, Johnson RS, Nizet V (2007) Cutting edge: Essential role of hypoxia inducible factor-1alpha in development of lipopolysaccharide-induced sepsis. J Immunol 178:7516–7519

Pollard JW (2004) Tumour-educated macrophages promote tumour progression and metastasis. Nat Rev Cancer 4:71–78

Portugal LR, Fernandes LR, Alvarez-Leite JI (2009) Host cholesterol and inflammation as common key regulators of toxoplasmosis and artherosclerosis development. Expert Rev Anti Infect Ther 7:807–819

Rius J, Guma M, Schachtrup C, Akassoglou K, Zinkernagel AS, Nizet V, Johnson RS, Haddad GG, Karin M (2008) NF-kappaB links innate immunity to the hypoxic response through transcriptional regulation of HIF-1alpha. Nature 453:807–811

Rydberg EK, Salomonsson L, Hulten LM, Noren K, Bondjers G, Wiklund O, Bjornheden T, Ohlsson BG (2003) Hypoxia increases 25-hydroxycholesterol-induced interleukin-8 protein secretion in human macrophages. Atherosclerosis 170:245–252

Sang N, Fang J, Srinivas V, Leshchinsky I, Caro J (2002) Carboxyl-terminal transactivation activity of hypoxia-inducible factor 1 alpha is governed by a von Hippel-Lindau protein-independent, hydroxylation-regulated association with p300/CBP. Mol Cell Biol 22:2984–2992

Semenza GL (2009) Regulation of oxygen homeostasis by hypoxia-inducible factor 1. Physiology (Bethesda) 24:97–106

Sica A, Schioppa T, Mantovani A, Allavena P (2006) Tumour-associated macrophages are a distinct M2 polarised population promoting tumour progression: potential targets of anti-cancer therapy. Eur J Cancer 42:717–727

Sivakumar B, Akhavani MA, Winlove CP, Taylor PC, Paleolog EM, Kang N (2008) Synovial hypoxia as a cause of tendon rupture in rheumatoid arthritis. J Hand Surg Am 33:49–58

Standiford TJ, Strieter RM, Lukacs NW, Kunkel SL (1995) Neutralization of IL-10 increases lethality in endotoxemia. Cooperative effects of macrophage inflammatory protein-2 and tumor necrosis factor. J Immunol 155:2222–2229

Szekanecz Z, Szegedi G, Koch AE (1998) Angiogenesis in rheumatoid arthritis: pathogenic and clinical significance. J Investig Med 46:27–41

Talks KL, Turley H, Gatter KC, Maxwell PH, Pugh CW, Ratcliffe PJ, Harris AL (2000) The expression and distribution of the hypoxia-inducible factors HIF-1alpha and HIF-2alpha in normal human tissues, cancers, and tumor-associated macrophages. Am J Pathol 157:411–421

Tracey KJ, Fong Y, Hesse DG, Manogue KR, Lee AT, Kuo GC, Lowry SF, Cerami A (1987) Anti-cachectin/TNF monoclonal antibodies prevent septic shock during lethal bacteraemia. Nature 330:662–664

Ulloa L, Tracey KJ (2005) The "cytokine profile": a code for sepsis. Trends Mol Med 11:56–63

Vaupel P, Kelleher DK, Hockel M (2001) Oxygen status of malignant tumors: pathogenesis of hypoxia and significance for tumor therapy. Semin Oncol 28:29–35

Walmsley SR, Cadwallader KA, Chilvers ER (2005a) The role of HIF-1alpha in myeloid cell inflammation. Trends Immunol 26:434–439

Walmsley SR, Print C, Farahi N, Peyssonnaux C, Johnson RS, Cramer T, Sobolewski A, Condliffe AM, Cowburn AS, Johnson N, Chilvers ER (2005b) Hypoxia-induced neutrophil survival is mediated by HIF-1alpha-dependent NF-kappaB activity. J Exp Med 201:105–115

White JR, Harris RA, Lee SR, Craigon MH, Binley K, Price T, Beard GL, Mundy CR, Naylor S (2004) Genetic amplification of the transcriptional response to hypoxia as a novel means of identifying regulators of angiogenesis. Genomics 83:1–8

Yu F, White SB, Zhao Q, Lee FS (2001) Dynamic, site-specific interaction of hypoxia-inducible factor-1alpha with the von Hippel-Lindau tumor suppressor protein. Cancer Res 61:4136–4142

Zemplenyi T, Crawford DW, Cole MA (1989) Adaptation to arterial wall hypoxia demonstrated in vivo with oxygen microcathodes. Atherosclerosis 76:173–179

Zhou J, Schmid T, Brune B (2003) Tumor necrosis factor-alpha causes accumulation of a ubiquitinated form of hypoxia inducible factor-1alpha through a nuclear factor-kappaB-dependent pathway. Mol Biol Cell 14:2216–2225

Zisman DA, Kunkel SL, Strieter RM, Gauldie J, Tsai WC, Bramson J, Wilkowski JM, Bucknell KA, Standiford TJ (1997) Anti-interleukin-12 therapy protects mice in lethal endotoxemia but impairs bacterial clearance in murine Escherichia coli peritoneal sepsis. Shock 8:349–356

Hypoxia and Metastasis in Breast Cancer

Helene Rundqvist and Randall S. Johnson

Contents

Abstract In this review we summarize the evidence for a role for hypoxic response in the biology of metastasis, with a particular emphasis on the metastasis of breast cancer and the function of the hypoxia inducible factor (HIF).

H. Rundqvist and R.S. Johnson (✉)
Molecular Biology Section, Division of Biological Sciences, University of California, San Diego, CA 92093, USA
e-mail: rjohnson@biomail.ucsd.edu

M. Celeste Simon (ed.), *Diverse Effects of Hypoxia on Tumor Progression*,
Current Topics in Microbiology and Immunology 345, DOI 10.1007/82_2010_77
© Springer-Verlag Berlin Heidelberg 2010, published online: 10 June 2010

1   Introduction

Hypoxic response in tumors is complex, and is an intrinsic aspect of virtually all solid tumor physiology. In many tumors, it may play only a peripheral role, particularly if the tumors remain small, nonnecrotic and nonmetastatic. In others, and especially in rapidly expanding and metastasizing masses, hypoxia and ischemia appear to be key elements in the biology and natural history of the disease. We will here review evidence that hypoxic response is a central component of metastasis and the metastatic process, particularly in human breast cancer and its animal models.

2   Tumors and Hypoxia

Growing solid tumors become hypoxic when blood supply is insufficient, either due to outgrowth of the existing vasculature, or as a result of formation of a dysfunctional vasculature. In response to hypoxia, tissues attempt to restore oxygen delivery by physiologic adjustment. That vascularization – necessary for the growth of the primary tumor and for metastasis – is driven by hypoxia and mediated mainly by activation of the hypoxia inducible transcription factors (HIFs) and expression of their target genes, e.g. vascular endothelial growth factor (VEGF) is relatively well understood. However, even though recruitment of blood and lymphatic vessels by secretion of pro-angiogenic factors enables migration of tumor cells from the primary tumor to distant metastatic sites, it does not provide a full explanation for why tumor hypoxia is a hallmark of increased malignant progression and diminished therapeutic response (Tatum et al. 2006). For example, in breast cancer, hypoxia is one of the most significant indicators of poor clinical outcome (Vaupel et al. 2002, 2005; Chaudary and Hill 2006).

2.1   *Hypoxia Inducible Factor Expression in Breast Cancer Is Highly Correlated with Metastasis and Mortality*

The members of the hypoxia inducible factor (HIF) transcription factor family are key mediators of adaptation to hypoxia, and are responsible for many of the features of hypoxia-driven tumor growth, especially with regard to angiogenic drive and metabolic adaptation. Increased HIF-1α expression is also closely linked to metastasis and poor prognosis in both hereditary and sporadic breast cancers (Bos et al. 2001, 2003; Schindl et al. 2002; Gruber et al. 2004; Dales et al. 2005; Generali et al. 2006; van der Groep et al. 2008). One of the first clinical studies of HIF-1α and breast cancer was carried out by Zhong et al., who reported that HIF-1α protein detected by immunohistochemistry is often highly expressed in breast tumors, as well as bordering "normal" areas adjacent to tumors (Zhong et al. 1999). A subsequent study by Bos et al. followed, which correlated the levels of HIF-1α over-expression with other prognostic factors of breast tumors, including proliferation rates, VEGF expression, microvessel density (MVD), expression of estrogen receptor (ER), and p53 expression (Bos et al. 2001). In line with the findings of Zhong et al., a majority of well-differentiated as well as poorly differentiated ductal carcinoma in situ (DCIS) tumors over-expressed HIF-1α, whereas HIF-1α was expressed in less than 1% of the normal breast epithelium, and was not detectable in normal nonepithelial breast tissue or ductal hyperplasias (Bos et al. 2001). HIF-1α over-expression was also noted in all tumors classified as poorly differentiated invasive carcinomas. Moreover, in DCIS lesions, HIF-1α over-expression was clearly associated with increased vessel density. Increased levels of HIF-1α protein were also associated with high levels of proliferation and increased expression of VEGF and ER in all breast tumor types (Bos et al. 2004). Furthermore, the C1772T polymorphism in the HIF-1α gene has been shown to increase the risk of developing breast cancer, nodal metastasis, and correlates with expression of HIF-1α in tumors (Kim et al. 2008; Naidu et al. 2009).

Overexpression of HIF-1 is clearly clinically relevant to survivability of breast cancer. It has now been shown in at least six studies of patient breast cancer biopsies that increased expression of HIF-1α is highly correlated with increased mortality and metastasis (Schindl et al. 2002; Bos et al. 2004; Currie et al. 2004; Gruber et al. 2004; Dales et al. 2005; Gao and Vande Woude 2005). This is true in both lymph node negative and lymph node positive primary breast tumors. Indeed, one study suggested that HIF-1α expression may be one of the best histological markers available for determination of poor prognosis (Dales et al. 2005).

3   The Metastatic Role of HIFs in Breast Cancer

3.1   Mammary Specific HIF Deletion in a Murine Model of Breast Cancer

In a mouse model of mammary carcinoma, deletion of HIF-1α in the mammary epithelium led to reduced tumor growth as well as decreased pulmonary metastasis (Liao et al. 2007). In addition, HIF-1 knock out MECs (mammary epithelial cells) show reduced proliferation, migration and invasion under hypoxia in vitro (Liao et al. 2007). Intense efforts are being made to identify molecular targets that mediate the pro-metastatic role of HIF, and there is emerging evidence that HIF expression directly promotes key factors for invasiveness, motility, and homing of the tumor cell to the metastatic niche.

3.2   Regulation of HIF Protein Levels Under Hypoxia

The HIF complex consists of one constitutively active β subunit and one of the hypoxia sensitive α subunits (1α or 2α) (Wang et al. 1995). Activation of HIF to a fully competent transcriptional regulatory protein complex is a multi-step process (Ruas and Poellinger 2005) that include protein stabilization and transactivation. At normal oxygen concentrations, HIF-α interacts with the ubiquitin protein ligase pVHL, promoting HIF-α degradation by the proteasome (Maxwell et al. 1999; Tanimoto et al. 2000). The interaction between pVHL and HIF-α is dependent on hydroxylation of two conserved proline residues (Ivan et al. 2001; Jaakkola et al. 2001). Hydroxylation is mediated by a family of prolyl hydroxylases (PHD) (Bruick and McKnight 2001; Epstein et al. 2001) which utilizes oxygen and 2-oxo-glutarate as substrates and generates CO_2 and succinate as by-products (Pan et al. 2007). Under hypoxic conditions, hydroxylation and the subsequent degradation are inhibited, resulting in accumulation of HIF-α protein. In addition to the protein stability, the transactivation capacity of HIF-α is regulated by factor inhibiting HIF (FIH) through hydroxylation of an asparagine residue within the transactivation domain (Mahon et al. 2001).

The notion that HIF activity is closely linked to poor prognosis in breast cancer is supported by the recent results by Yan et al., showing that in aggressive familial breast cancers linked to BRCA1 mutation there is suppressed expression of PHDs and atypical localization of FIH (Yan et al. 2009), the two negative regulators of HIF. In addition to low oxygen availability, alterations in abundance and activity of PHDs modulate HIF actions both in vivo and in vitro (Pan et al. 2007; Aragones et al. 2008; Tennant et al. 2009). The tricarboxylic acid (TCA) cycle metabolite succinate is a potent inhibitor of PHD activity. Succinate accumulates and translocates from the mitochondria to the cytoplasm when the TCA enzyme succinate dehydrogenase (SDH) is inhibited. Reduced SDH activity has been shown in cells from human mammary carcinomas (Putignani et al. 2008) and accumulation of succinate has been shown to modulate HIF activity through downregulation of PHD activity (Selak et al. 2005). This links the metabolic perturburation primarily caused by HIF in cancer cells to a further activation of the HIF system. Furthermore, cytoplasmic FIH expression correlates to poor prognosis in invasive breast cancer (Tan et al. 2007). Taken together these results indicate that moderation of the HIF regulatory system may contribute to the hypoxic phenotype and aggressiveness of certain breast cancers.

3.3   Differential and Overlapping Features of HIF-1 and HIF-2

HIF-1 and HIF-2 regulate similar sets of genes, however, much data argue that HIF-1 is expressed in most tissues and transiently activated in response to hypoxia, while HIF-2 is expressed in distinct cell populations and responds to less severe and chronic hypoxic exposure (Poellinger and Johnson 2004; Holmquist-Mengelbier et al. 2006). Studies have shown that in the context of loss of HIF degradation, HIF-2α becomes the dominant HIF-α isoform (Rankin et al. 2008). Like HIF-1α, HIF-2α is overexpressed in a number of human cancers. In VHL deficient renal carcinomas, HIF-2α seems to be largely responsible for tumor progression, for review see Qing and Simon (2009). In human breast carcinomas, HIF-2α expression correlates to metastatic ability (Giatromanolaki et al. 2006).

3.4   HIF Regulated Angiogenic Factors and Metastasis

HIFs are key mediators of angiogenesis. HIF-1 and HIF-2 regulate the expression of an orchestra of pro-angiogenic factors such as VEGF, VEGFR1, angiopoietin and erythropoietin (Semenza 2003). The excess of pro-angiogenic signaling leads to vascular abnormality. The leaky, tortuous vessels in growing solid tumors lead to poor perfusion, an increase in hypoxic area and may facilitate intravasation of tumor cells (Jain 2005). Modulation of pro-angiogenic stimuli can normalize the vasculature (Jain 2005; Dickson et al. 2007; Stockmann et al. 2008; Heath and Bicknell 2009; Hedlund et al. 2009; Mazzone et al. 2009) and possibly reduce metastasis (Mazzone et al. 2009).

There is further evidence that the characteristic of the vascular microenviron-ment is an important factor for tumor cell function/invasiveness. Depletion of HIF-1α in transformed astrocytes led to reduced growth and angiogenesis compared to controls when injected subcutaneously, where there is low inherent vasculature (Blouw et al. 2003). In contrast, when the HIF-1α depleted cells were introduced to the highly vasculated brain parenchyma, tumors grew faster and spread to both hemispheres more rapidly than control cells (Blouw et al. 2003). These results indicate that HIF-1α-deficient astrocytomas were unable to induce neo-angiogenesis, but adapted to the milieu and became more motile and invasive, moving along the existing vasculature.

VEGF per se may induce invasion: in a study by Cannito et al., incubation of a breast cancer cell line with VEGF-containing media, as well as media from hypoxic exposure of the same cell line, both induced invasion as measured by a matrigel/Boyden assay (Cannito et al. 2008). The results indicate an auto/paracrine effect of VEGF on breast cancer cell invasiveness. Similar results have been obtained in models of pancreatic and prostate cancer (Gonzalez-Moreno et al. 2010; Yang et al. 2006), where VEGF-stimulated cells acquire epithelial to mesenchymal transition (EMT) features along with increases in motility and invasiveness. These studies describe a nonangiogenic effect of VEGF that may play a role in early tumor spreading.

VEGF may also indirectly promote metastasis by its ability to induce actin rearrangement in endothelial cells and increase vascular permeability. This will supposedly facilitate metastasis in two ways, by causing gaps between endothelial cells and thereby aiding intravasation of tumor cells, and also by altering the interstitial fluid pressure. Tumors with high interstitial fluid pressure have been associated with increased metastasis and lower disease free survival (Milosevic et al. 2001; Rofstad et al. 2002).

Furthermore, VEGF is also involved in formation of the premetastatic niche by recruitment of VEGFR1 positive myeloid cells to the premetastatic site (Kaplan et al. 2005), this finding has, however, been challenged by recent results (Dawson et al. 2009) where blocking of VEGFR1 did not decrease metastatic burden.

3.5   A Link Between Tumor Metabolism/Acidosis and Metastasis

A common feature of hypoxic cells is that they depend on anaerobic metabolism for their ATP production. This is called the Pasteur effect and is mediated by HIF-1 (Seagroves et al. 2001). This enhanced glycolysis will lead to an increase in lactate production and acidosis. Acidosis per se is known to promote tumor cell invasion by destruction of adjacent tissue and degradation of the extracellular matrix (ECM) (Martinez-Zaguilan et al. 1996). In addition, acidosis can reduce the effect of cancer drugs that work optimally only under normal pH.

Intracellular lactate needs to be extruded to allow for sustained glycolysis and ATP production. Lactate is cleared by export to the ECM by H^+/lactate co-transporters such as the HIF-regulated monocarboxylate transporter 4 (MCT4) (Ullah et al. 2006). Excess lactate can be oxidized in better-oxygenated tumor cells (Sonveaux 2008) as well as in stromal cells (Koukourakis et al. 2006). In addition to lactate production, the high overall metabolic rate in tumors produces significant amounts of CO_2 that further contribute to acidosis. The enzyme carbonic anhydrase IX (CAIX) is located in the plasma membrane and converts CO_2 and water to HCO_3^- and H^+. HCO_3^- re-enters the cell and aid in intracellular alkalinization. As a result, despite the high rate of anaerobic metabolism, the intracellular pH of tumors are kept in the physiological range of 7.0–7.4, while the extracellular pH can be as low as 6.0–7.0 (Vaupel et al. 1990). HIF-1 is responsible for regulation of CAIX expression and CAIX has been widely used as a marker of hypoxia.

CAIX overexpression has been shown in a number of tumors, such as cervix, breast, and lung carcinomas and is related to prognosis, both linked to HIF-1 and independently (Giatromanolaki et al. 2001; Bartosova et al. 2002). In a study by Chia et al., CAIX was associated with negative ER status, higher relapse rate and worse overall survival in breast cancer patients (Chia et al. 2001).

3.5.1   Therapeutic Targets of Metabolic Disease in Breast Cancer Progression

Glucose dependent metabolism has been considered a possible target for cancer therapy in combination with conventional treatment, since it targets cells highly resistant to chemo- and radiotherapy (Scatena et al. 2008). Use of the metabolic drug metformin in diabetic patients decreases breast cancer incidence and mortality (Evans et al. 2005; Jiralerspong et al. 2009) and the AMP analog AICAR that mimics low energy status in the cell has been shown to reduce growth and invasion in a breast cancer model (Swinnen et al. 2005).

Mechanistically, both Metformin and AICAR induce AMPK activity. AMPK activation per se can induce cell cycle arrest, downregulate mTOR activity and reduce HER-2 expression (Dowling et al. 2007; Zhuang and Miskimins 2008; Liu et al. 2009; Oliveras-Ferraros et al. 2009; Phoenix et al. 2009; Vazquez-Martin et al. 2009). Systemically, AMPK activation increases glucose metabolism and reduces circulating insulin and IGF-1 levels in patients with hyperinsulinemia. Treatment with Metformin and AICAR have been shown to reduce blood insulin levels also in breast cancer patients and many breast cancers express insulin receptors as well as receptors for IGF-1 (Goodwin et al. 2008). In addition to their mitogenic effect as growth factors, insulin and IGF-1 have been shown to induce HIF activity (Zelzer et al. 1998). Treating epithelial cells with AICAR or metformin inhibits the ability of insulin and IGF-1 to induce HIF-1α expression (Treins et al. 2006).

In line with this, both Metformin and AICAR reduce HIF-1 activity in epithelial cells (Treins et al. 2006). This may occur through the inhibitory effect of AMPK on mTOR, since mTOR and the PI3K/Akt pathway are involved in regulation of HIF in breast cancer cells (Blancher et al. 2001). Taken together, reduced HIF activity may contribute to the beneficial effects of metformin in treatment of breast cancer, as hinted at in Shackelford et al. (2009).

However, AMPK activation is capable of inducing HIF-1-independent VEGF production (Yun et al. 2005). This introduces a possible negative effect of metformin treatment that is supported by the recent reports where metformin appear to increase VEGF expression and induce angiogenesis in a breast cancer model (Phoenix et al. 2009).

3.6   Motility and the Stroma

3.6.1   The Invasiveness of a Tumor Cell Depends on Its Ability to Migrate Through the Stroma

The invasiveness of a tumor cell depends on its ability to migrate through the stroma, but also on breakdown of the ECM and entry into the vasculature. The migratory phenotype requires changes in src, rac-1, Rho activation. Loss of HIF-1 in MECs reduces directed motility as well as invasion under hypoxia (Liao et al. 2007). Several HIF target genes have been shown to contribute to this phenotype.

The hepatocyte growth factor (HGF) pathway is associated with proliferation, motility, invasion and angiogenesis in breast cancer and enhances the transition from DCIS to invasive carcinoma (Gao and Vande Woude 2005; Jedeszko et al. 2009). HGF is expressed by fibroblasts in the tumor stroma and secreted as a precursor form (proHGF). It is then converted to active HGF by the proteinase hepatocyte growth factor activator (HGFA) and act through its receptor c-MET

(Hanna et al. 2009). c-MET is overexpressed on aggressive forms of breast cancer and associated with invasion and metastasis (Lee et al. 2005). Expression and activation of c-MET has been correlated to tumor hypoxia and is promoted by HIF-1 (Pennacchietti et al. 2003; Chen et al. 2007). In addition to regulating the expression of c-MET, HIF-1 has also been suggested to induce expression of HGFA in tumor cells and thereby increase the bioavailable levels of HGF in the stroma and further activate c-MET signaling (Kitajima et al. 2008) and metastatic potential.

Macrophage-stimulating protein (MSP) is a HGF homologue that promotes metastasis in animal models (Zinser et al. 2006). It is the only known ligand of Recepteur d'origine nantais (RON), a predictor of metastasis and low survival in breast cancer patients (Lee et al. 2005). RON was recently shown to be a direct target of HIF-1 (Thangasamy et al. 2009).

A transient hypoxia exposure prior to intravenous injection increases the metastatic potential of several cancer cell lines (Young et al. 1988; Young and Hill 1990), indicating that hypoxic exposure also contributes to cancer cell extravasation and settlement. The HIF-1 regulated chemokine receptor CXCR4 (Schioppa et al. 2003; Staller et al. 2003) is highly prevalent on breast cancer cells and responsible for cell migration but also recruitment of tumor cells to the premetastatic niche (Muller et al. 2001). Its ligand, stromal derived factor-1 (SDF-1/CXCL12), is expressed by cancer cells, normal MECs, and fibroblasts in the tumor stroma (Orimo et al. 2005; Du et al. 2008; Serrati et al. 2008). In a glioblastoma model, it was shown that hypoxia and HIF-1 induced tumor cell expression of SDF-1 and increased recruitment of CXCR4 positive BMDCs to the tumor stroma (Du et al. 2008). HIF-1 mediated expression of SDF-1 has also been reported in inflammatory fibroblasts and led to infiltration of myeloid cells (del Rey et al. 2009). The recruitment of BMDCs is important for vascularization of the primary tumor (Du et al. 2008) but may also play a role in the formation of metastases. Recently, Dunn et al., showed that high levels of TGF and HIF-1 in the bone microenvironment up-regulate local CXCR4 and VEGF expression and promote metastasis (Dunn et al. 2009).

Recently, another chemokine, CXCR6, was shown to be regulated by HIF-1 and induce migration in breast cancer cells. CXCR6 is highly expressed in breast cancer cells in vitro and in metastatic lymph nodes, but not to the same extent in the primary tumor (Lin et al. 2009). The authors show a chemotactic effect of the ligand CXCL16 on CXCR6 positive breast cancer cells and suggest it as a possible way to recruit tumor cells to the lymph nodes that are known to express the CXCL16.

When screening for genes associated with hypoxia in transformed cells, the amine oxidase LOX was identified (Denko et al. 2003). LOX was previously known to crosslink collagens and elastins in the ECM but had also been associated with increased breast cancer cell invasion in vitro (Kirschmann et al. 2002). Erler et al., showed that high expression of LOX is correlated with poor prognosis in ER negative breast cancer patients (Erler et al. 2006) and in the same publication they show that LOX has an HRE in its promoter region and is regulated by HIF-1. Inhibition of LOX reduces hypoxia induced cancer cell motility and invasiveness and prevents metastasis in vivo (Erler et al. 2006). The molecular mechanisms behind the metastatic potential of LOX include activation of focal adhesion kinase activity (Erler et al. 2006), stabilization of SNAIL (Sahlgren et al. 2008) and downregulation of E-cadherin and induction of EMT (Schietke et al. 2010). Furthermore, secreted LOX has been shown to accumulate at premetastatic sites and initiate formation of the premetastatic niche by crosslinking collagen IV and thereby induce recruitment of first myeloid cells and then bone marrow-derived cells and tumor cells (Erler et al. 2009). It was recently shown that inhibitors of LOX have no effect on already established metastasis, supporting the notion that LOX activity is important primarily in the early metastatic process (Bondareva et al. 2009).

3.6.2 Epithelial to Mesenchymal Transition

Oxygen availability also regulates EMT, a dedifferentiation process where the tumor cells lose expression of cell adhesion molecules such as E-cadherin, and tight epithelial cell–cell contacts are dissolved. The cell acquires a spindle-like, mesenchymal, migratory phenotype and becomes increasingly independent of ECM and stromal cells for survival (Thiery 2002; Grunert et al. 2003) leading to enhanced invasion and metastasis. Although EMT is induced by a number of stimuli such as HGF, TGFβ, FGF, PDGF and Wnt ligands, it is now widely accepted that hypoxia alone can induce EMT and a consequent increase in invasiveness also in breast cancer cells (Lester et al. 2007; Cannito et al. 2008). E-cadherin is one of the main adhesion molecules in the epithelium and loss of e-cadherin is regarded as a central event in metastasis, and is closely linked to hypoxia and HIF expression (Esteban et al. 2006). Inhibition of E-cadherin in vitro leads to increased invasiveness, and over-expression of E-cadherin in already invasive cells can reduce invasiveness (Hanahan and Weinberg 2000). Multiple mechanisms contribute to the inactivation of E-cadherin, including promoter methylation, phosphorylation, and transcriptional repression.

HIF-1 regulates several transcription factors/repressors involved in EMT; e.g., SNAIL, SLUG, TWIST, and ZEB1 (for review see Haase 2009). These factors regulate migration and invasion in hypoxic breast cancer cells, partly by repressing E-cadherin expression through interaction with e-boxes located in the proximal promoter of E-cadherin (Chen et al. 2010; Yang et al. 2004; Eger et al. 2005; Alexander et al. 2006).

In addition to losing adhesion molecules, cancer cells that undergo EMT also acquire new ways to physically interact with the microenvironment. Loss of E-cadherin is often accompanied with activation of mesenchymal markers, e.g., N-cadherin and vimentin. In melanoma, this cadherin switch can be a direct effect, as shown by Kuphal et al., where expression of the intracellular domain of E-cadherin was sufficient to down-regulate N-cadherin expression and invasiveness, possibly mediated by another hypoxia-sensitive transcription factor,

NFκB (Kuphal and Bosserhoff 2006). In addition, the HIF regulated factor TWIST can down-regulate E-cadherin but also activate N-cadherin (Yang et al. 2004). Transfection of N-cadherin to a weakly metastatic breast cancer cell line promoted motility, invasion, and metastasis (Hazan et al. 2000). Cadherins usually form homotypic aggregation, but N-cadherin can also form heterotypic adhesions to other cell types in the stroma, such as endothelial cells and fibroblasts (Li et al. 2001). The HIF-1 regulated $\alpha5\beta1$ integrin is expressed in breast cancer cells in response to hypoxia and provides adhesion to fibronectin and laminin; binding promotes cell survival and mobility through ECM and on a fibroblasts layer (Spangenberg 2006). $\alpha5\beta1$ ligation has also been linked to survival of metastatic breast cancer cells in bone marrow metastasis (Korah et al. 2004).

Some HIF-1 regulated metabolic genes are also involved in promoting motility. For example, MCT4 interacts with $\beta1$ integrins in lammelapodia inducing migratory ability (Gallagher et al. 2009) and phosphoglucose isomerase (PGI, aka autocrine motility factor, or AMF) is known to stimulate cell motility (Niizeki et al. 2002; Funasaka et al. 2005). Recent evidence also showed that overexpression of PGI can lead to EMT in breast cancer cells (Funasaka et al. 2009). This is supported by the results of Tsutsumi et al., showing that AMF can induce SNAIL and reduce E-cadherin expression (Tsutsumi et al. 2004).

3.6.3 Stromal Interaction and ECM Degradation

Acquisition of the ability to break down the ECM and basement membrane surrounding the primary tumor is part of the cancer cells path to becoming invasive, and also seems to be an important trait for extravasation and metastasis formation. HIFs are part of this path, and can activate proteolytic enzymes such as MMP-2, MMP-9, MMP-14, and Cathepsin D, either directly or indirectly (Petrella et al. 2005; Huang et al. 2009; Jo et al. 2009; Song et al. 2009). These enzymes have no proven effect on migratory ability, but alter the vascular architecture (Chabottaux et al. 2009) and aid invasiveness and settlement in the metastatic tissue (Rizki et al. 2008; Huang et al. 2009).

The tumor associated proteolytic factor uPA/uPAR was early on shown to be regulated by hypoxia and important for invasion (Graham et al. 1998, 1999; Kroon et al. 2000; Maity and Solomon 2000). The uPA becomes active when bound to the HIF regulated receptor uPAR, the complex then converts plasminogen into plasmin (Graham et al. 1998; Buchler et al. 2009). Plasmin contributes to proteolytic activity both by direct degradation of ECM proteins and by activating MMPs. Several studies have found that high levels of the uPA system is associated with tumor aggressiveness and poor prognosis in breast cancer (Harbeck et al. 2007). Interestingly, the expression is equally predictive coming from tumor cells as well as the tumor stroma (Hildenbrand and Schaaf 2009). Recent findings also indicate that uPAR may contribute to EMT in cancer cells (Jo et al. 2009).

4 Alternative Regulation of HIF

4.1 Oncogene Signaling Alters HIF Activity

In addition to hypoxia, deregulation of oncogenes and tumor supressor genes can alter HIF activity, primarily through the PI3K/Akt, mTOR, NFκB and Notch signaling pathways. For example, HER-2/neu overexpression leads to increased HIF-1 stability through mTOR (Semenza 2003). Notch signaling has been shown to mediate hypoxic activation of EMT and directly upregulate Snail1 and Slug (Leong et al. 2007; Sahlgren et al. 2008) but with HIF-1α as part of the transcriptional complex. Inhibition of Notch in this model prevented E-cadherin downregulation and conserved the epithelial morphology under hypoxia. In addition, Notch activation augmented HIF-1 recruitment to the LOX promoter, LOX mRNA expression and the subsequent increase in protein levels of Snail1 under hypoxic conditions (Sahlgren et al. 2008). Recently, it has been shown that HIF potentiated Notch signaling and downstream e-cadherin repression in breast cancer cell lines (Chen et al. 2010).

NFκB-HIF-1 interaction has been shown to contribute to EMT and breast cancer metastatic capacity (Bendinelli et al. 2009). Functional NFκB is also necessary for HIF-1α mRNA expression in macrophages and fibroblasts (Belaiba et al. 2007; Rius et al. 2008), the two most prominent cell types of the tumor stroma. Both macrophages and fibroblasts significantly contribute to breast cancer metastasis (Lin et al. 2001; Liao et al. 2009) and, as previously mentioned, macrophage infiltration correlates to poor prognosis in breast cancer (Leek et al. 1996; Tsutsui et al. 2005).

4.2 Dual Roles of Nitric Oxide in Tumor Progression

Nitric oxide has been shown to have both tumor promoting roles such as increased vascular permeability, blood flow, MMP expression and angiogenesis, but also anti-tumor roles such as induced tumor cell death. NO is one of the key signaling molecules in inflammation and it modulates HIF-1 activity in normoxia as well as hypoxia (for a review see Berchner-Pfannschmidt et al. 2010). Macrophage derived NO has been suggested to kill tumor cells by apoptosis, but with a subsequent shift in macrophage phenotype towards the M2 polarization state, reviewed in Weigert and Brune (2008). One characteristic of M2 macrophages is upregulated arginase expression and activity. Arginase competes with iNOS for their common substrate ι-arginine and in contrast to M1 polarized macrophages, M2 polarized macrophages produce low levels of NO upon stimulation (Rauh et al. 2005). It has recently been shown that the HIF-1/HIF-2 balance is part of the divergence in NO production between M1 and M2, where M2 macrophages primarily express HIF-2α (Takeda et al. 2010). In the study by Takeda et al., HIF-2 induces arginase production while

M1 macrophages have a HIF-1 profile that increase iNOS expression. Expression of iNOS in tumor cells is associated with apoptosis, suppression of tumorigenicity, abrogation of metastasis and regression of established hepatic metastases (see the review by Lechner et al. (2005) and Tatemichi et al. (2009)). In addition, macrophage NO production in tumors has been shown to have a cytotoxic and cytostatic effect (Hibbs et al. 1987; Xie et al. 1996). Together these data suggests that reduced NO production may be part of how tumor associated macrophages facilitate tumor progression.

4.3   Deacetylation Inhibitors in HIF Regulation

A recent study indicates a possible modulation of NFκB and HIF-1 activity in breast cancer cells by histone deacetylases (HDACs) (Bendinelli et al. 2009). HDACs regulate transcriptional activity by removing lysine bound acetyl groups. For example, they repress transcription by deacetylation of histone tails and thereby increase histone tail affinity to DNA and induce gene silencing. HDAC expression and activity has been shown to increase in response to hypoxia (Kim et al. 2001). Activation of HDAC1 is associated with an angiogenic profile in several cell types, and has been shown to down-regulate pVHL expression but increase the expression of HIF-1α and VEGF (Kim et al. 2001). Global inhibition of deacetylases results in apoptosis of cancer cells and reduced tumor growth, and several HDAC inhibitors are currently being tested in clinical trials (for a review see Mottet and Castronovo (2008)). A decrease in HIF-1α levels was recently suggested as the main explanation for reduced formation of metastatic lung lesions by the HDAC inhibitor AN-7 in a breast carcinoma model (Tarasenko et al. 2008).

5   Concluding Remarks

Tumor cells are educated in the hypoxic environment of the primary tumor and there gain features necessary for the malignant progression and establishment of metastatic tumors. How hypoxia impacts this process across the wide range of cells involved in tumorigenesis is a key question that remains to be answered. It is also a question that requires the synthesis of a great deal of data studying the individual influence of the hypoxic response in each cell type. Interestingly, data indicate that the process of breast cancer metastasis, and metastasis generally, is impacted by hypoxic response in all of the cell types involved in the metastatic process, e.g., malignant cells, endothelial cells, inflammatory cells and mesenchymal cell types. Although synthesizing a complete picture of the role of hypoxic response in the metastatic process will involve studying all of these, this should allow important targets for interrupting metastasis to be discovered and evaluated.

References

Alexander NR, Tran NL et al (2006) N-cadherin gene expression in prostate carcinoma is modulated by integrin-dependent nuclear translocation of Twist1. Cancer Res 66(7):3365–3369

Aragones J, Schneider M et al (2008) Deficiency or inhibition of oxygen sensor Phd1 induces hypoxia tolerance by reprogramming basal metabolism. Nat Genet 40(2):170–180

Bartosova M, Parkkila S et al (2002) Expression of carbonic anhydrase IX in breast is associated with malignant tissues and is related to overexpression of c-erbB2. J Pathol 197(3):314–321

Belaiba RS, Bonello S et al (2007) Hypoxia up-regulates hypoxia-inducible factor-1alpha transcription by involving phosphatidylinositol 3-kinase and nuclear factor kappaB in pulmonary artery smooth muscle cells. Mol Biol Cell 18(12):4691–4697

Bendinelli P, Matteucci E et al (2009) NF-kappaB activation, dependent on acetylation/deacetylation, contributes to HIF-1 activity and migration of bone metastatic breast carcinoma cells. Mol Cancer Res 7(8):1328–1341

Berchner-Pfannschmidt U, Tug S et al (2010) Oxygen-sensing under the influence of nitric oxide. Cell Signal 22(3):349–356

Blancher C, Moore JW et al (2001) Effects of ras and von Hippel-Lindau (VHL) gene mutations on hypoxia-inducible factor (HIF)-1alpha, HIF-2alpha, and vascular endothelial growth factor expression and their regulation by the phosphatidylinositol 3′-kinase/Akt signaling pathway. Cancer Res 61(19):7349–7355

Blouw B, Song H et al (2003) The hypoxic response of tumors is dependent on their microenvironment. Cancer Cell 4(2):133–146

Bondareva A, Downey CM et al (2009) The lysyl oxidase inhibitor, beta-aminopropionitrile, diminishes the metastatic colonization potential of circulating breast cancer cells. PLoS One 4 (5):e5620

Bos R, Zhong H et al (2001) Levels of hypoxia-inducible factor-1 alpha during breast carcinogenesis. J Natl Cancer Inst 93(4):309–314

Bos R, van der Groep P et al (2003) Levels of hypoxia-inducible factor-1alpha independently predict prognosis in patients with lymph node negative breast carcinoma. Cancer 97(6): 1573–1581

Bos R, van Diest PJ et al (2004) Expression of hypoxia-inducible factor-1alpha and cell cycle proteins in invasive breast cancer are estrogen receptor related. Breast Cancer Res 6(4): R450–R459

Bruick RK, McKnight SL (2001) A conserved family of prolyl-4-hydroxylases that modify HIF. Science 294(5545):1337–1340

Buchler P, Reber HA et al (2009) Transcriptional regulation of urokinase-type plasminogen activator receptor by hypoxia-inducible factor 1 is crucial for invasion of pancreatic and liver cancer. Neoplasia 11(2):196–206

Cannito S, Novo E et al (2008) Redox mechanisms switch on hypoxia-dependent epithelial-mesenchymal transition in cancer cells. Carcinogenesis 29(12):2267–2278

Chabottaux V, Ricaud S et al (2009) Membrane-type 4 matrix metalloproteinase (MT4-MMP) induces lung metastasis by alteration of primary breast tumor vascular architecture. J Cell Mol Med 13(9B):4002–4013

Chaudary N, Hill RP (2006) Hypoxia and metastasis in breast cancer. Breast Dis 26:55–64

Chen HH, Su WC et al (2007) Hypoxia-inducible factor-1alpha correlates with MET and metastasis in node-negative breast cancer. Breast Cancer Res Treat 103(2):167–175

Chen J, Imanaka N et al (2010) Hypoxia potentiates notch signaling in breast cancer leading to decreased E-cadherin expression and increased cell migration and invasion. Br J Cancer 102(2):351–360

Chia SK, Wykoff CC et al (2001) Prognostic significance of a novel hypoxia-regulated marker, carbonic anhydrase IX, in invasive breast carcinoma. J Clin Oncol 19(16):3660–3668

Currie MJ, Hanrahan V et al (2004) Expression of vascular endothelial growth factor D is associated with hypoxia inducible factor (HIF-1alpha) and the HIF-1alpha target gene

DEC1, but not lymph node metastasis in primary human breast carcinomas. J Clin Pathol 57(8):829–834

Dales JP, Garcia S et al (2005) Overexpression of hypoxia-inducible factor HIF-1alpha predicts early relapse in breast cancer: retrospective study in a series of 745 patients. Int J Cancer 116(5):734–739

Dawson MR, Duda DG et al (2009) VEGFR1-activity-independent metastasis formation. Nature 461(7262):E4; discussion E5

del Rey MJ, Izquierdo E et al (2009) Human inflammatory synovial fibroblasts induce enhanced myeloid cell recruitment and angiogenesis through a hypoxia-inducible transcription factor 1alpha/vascular endothelial growth factor-mediated pathway in immunodeficient mice. Arthritis Rheum 60(10):2926–2934

Denko NC, Fontana LA et al (2003) Investigating hypoxic tumor physiology through gene expression patterns. Oncogene 22(37):5907–5914

Dickson PV, Hamner JB et al (2007) Bevacizumab-induced transient remodeling of the vasculature in neuroblastoma xenografts results in improved delivery and efficacy of systemically administered chemotherapy. Clin Cancer Res 13(13):3942–3950

Dowling RJ, Zakikhani M et al (2007) Metformin inhibits mammalian target of rapamycin-dependent translation initiation in breast cancer cells. Cancer Res 67(22):10804–10812

Du R, Lu KV et al (2008) HIF1alpha induces the recruitment of bone marrow-derived vascular modulatory cells to regulate tumor angiogenesis and invasion. Cancer Cell 13(3):206–220

Dunn LK, Mohammad KS et al (2009) Hypoxia and TGF-beta drive breast cancer bone metastases through parallel signaling pathways in tumor cells and the bone microenvironment. PLoS One 4(9):e6896

Eger A, Aigner K et al (2005) DeltaEF1 is a transcriptional repressor of E-cadherin and regulates epithelial plasticity in breast cancer cells. Oncogene 24(14):2375–2385

Epstein AC, Gleadle JM et al (2001) C. elegans EGL-9 and mammalian homologs define a family of dioxygenases that regulate HIF by prolyl hydroxylation. Cell 107(1):43–54

Erler JT, Bennewith KL et al (2006) Lysyl oxidase is essential for hypoxia-induced metastasis. Nature 440(7088):1222–1226

Erler JT, Bennewith KL et al (2009) Hypoxia-induced lysyl oxidase is a critical mediator of bone marrow cell recruitment to form the premetastatic niche. Cancer Cell 15(1):35–44

Esteban MA, Tran MG et al (2006) Regulation of E-cadherin expression by VHL and hypoxia-inducible factor. Cancer Res 66(7):3567–3575

Evans JM, Donnelly LA et al (2005) Metformin and reduced risk of cancer in diabetic patients. BMJ 330(7503):1304–1305

Funasaka T, Yanagawa T et al (2005) Regulation of phosphoglucose isomerase/autocrine motility factor expression by hypoxia. FASEB J 19(11):1422–1430

Funasaka T, Hogan V et al (2009) Phosphoglucose isomerase/autocrine motility factor mediates epithelial and mesenchymal phenotype conversions in breast cancer. Cancer Res 69(13):5349–5356

Gallagher SM, Castorino JJ et al (2009) Interaction of monocarboxylate transporter C4 with beta1-integrin and its role in cell migration. Am J Physiol Cell Physiol 296(3):C414–C421

Gao CF, Vande Woude GF (2005) HGF/SF-Met signaling in tumor progression. Cell Res 15(1):49–51

Generali D, Berruti A et al (2006) Hypoxia-inducible factor-1alpha expression predicts a poor response to primary chemoendocrine therapy and disease-free survival in primary human breast cancer. Clin Cancer Res 12(15):4562–4568

Giatromanolaki A, Koukourakis MI et al (2001) Expression of hypoxia-inducible carbonic anhydrase-9 relates to angiogenic pathways and independently to poor outcome in non-small cell lung cancer. Cancer Res 61(21):7992–7998

Giatromanolaki A, Sivridis E et al (2006) Hypoxia-inducible factor-2 alpha (HIF-2 alpha) induces angiogenesis in breast carcinomas. Appl Immunohistochem Mol Morphol 14(1):78–82

Gonzalez-Moreno O, Lecanda J et al (2010) VEGF elicits epithelial-mesenchymal transition (EMT) in prostate intraepithelial neoplasia (PIN)-like cells via an autocrine loop. Exp Cell Res 316(4):554–567

Goodwin PJ, Pritchard KI et al (2008) Insulin-lowering effects of metformin in women with early breast cancer. Clin Breast Cancer 8(6):501–505

Graham CH, Fitzpatrick TE et al (1998) Hypoxia stimulates urokinase receptor expression through a heme protein-dependent pathway. Blood 91(9):3300–3307

Graham CH, Forsdike J et al (1999) Hypoxia-mediated stimulation of carcinoma cell invasiveness via upregulation of urokinase receptor expression. Int J Cancer 80(4):617–623

Gruber G, Greiner RH et al (2004) Hypoxia-inducible factor 1 alpha in high-risk breast cancer: an independent prognostic parameter? Breast Cancer Res 6(3):R191–R198

Grunert S, Jechlinger M et al (2003) Diverse cellular and molecular mechanisms contribute to epithelial plasticity and metastasis. Nat Rev Mol Cell Biol 4(8):657–665

Haase VH (2009) Oxygen regulates epithelial-to-mesenchymal transition: insights into molecular mechanisms and relevance to disease. Kidney Int 76(5):492–499

Hanahan D, Weinberg RA (2000) The hallmarks of cancer. Cell 100(1):57–70

Hanna JA, Bordeaux J et al (2009) The function, proteolytic processing, and histopathology of Met in cancer. Adv Cancer Res 103:1–23

Harbeck N, Schmitt M et al (2007) Tumor-associated proteolytic factors uPA and PAI-1: critical appraisal of their clinical relevance in breast cancer and their integration into decision-support algorithms. Crit Rev Clin Lab Sci 44(2):179–201

Hazan RB, Phillips GR et al (2000) Exogenous expression of N-cadherin in breast cancer cells induces cell migration, invasion, and metastasis. J Cell Biol 148(4):779–790

Heath VL, Bicknell R (2009) Anticancer strategies involving the vasculature. Nat Rev Clin Oncol 6(7):395–404

Hedlund EM, Hosaka K et al (2009) Malignant cell-derived PlGF promotes normalization and remodeling of the tumor vasculature. Proc Natl Acad Sci USA 106(41):17505–17510

Hibbs JB Jr, Taintor RR et al (1987) Macrophage cytotoxicity: role for L-arginine deiminase and imino nitrogen oxidation to nitrite. Science 235(4787):473–476

Hildenbrand R, Schaaf A (2009) The urokinase-system in tumor tissue stroma of the breast and breast cancer cell invasion. Int J Oncol 34(1):15–23

Holmquist-Mengelbier L, Fredlund E et al (2006) Recruitment of HIF-1alpha and HIF-2alpha to common target genes is differentially regulated in neuroblastoma: HIF-2alpha promotes an aggressive phenotype. Cancer Cell 10(5):413–423

Huang CH, Yang WH et al (2009) Regulation of membrane-type 4 matrix metalloproteinase by SLUG contributes to hypoxia-mediated metastasis. Neoplasia 11(12):1371–1382

Ivan M, Kondo K et al (2001) HIFalpha targeted for VHL-mediated destruction by proline hydroxylation: implications for O2 sensing. Science 292(5516):464–468

Jaakkola P, Mole DR et al (2001) Targeting of HIF-alpha to the von Hippel-Lindau ubiquitylation complex by O2-regulated prolyl hydroxylation. Science 292(5516):468–472

Jain RK (2005) Normalization of tumor vasculature: an emerging concept in antiangiogenic therapy. Science 307(5706):58–62

Jedeszko C, Victor BC et al (2009) Fibroblast hepatocyte growth factor promotes invasion of human mammary ductal carcinoma in situ. Cancer Res 69(23):9148–9155

Jiralerspong S, Palla SL et al (2009) Metformin and pathologic complete responses to neoadjuvant chemotherapy in diabetic patients with breast cancer. J Clin Oncol 27(20):3297–3302

Jo M, Lester RD et al (2009) Reversibility of epithelial-mesenchymal transition (EMT) induced in breast cancer cells by activation of urokinase receptor-dependent cell signaling. J Biol Chem 284(34):22825–22833

Kaplan RN, Riba RD et al (2005) VEGFR1-positive haematopoietic bone marrow progenitors initiate the pre-metastatic niche. Nature 438(7069):820–827

Kim MS, Kwon HJ et al (2001) Histone deacetylases induce angiogenesis by negative regulation of tumor suppressor genes. Nat Med 7(4):437–443

Kim HO, Jo YH et al (2008) The C1772T genetic polymorphism in human HIF-1alpha gene associates with expression of HIF-1alpha protein in breast cancer. Oncol Rep 20(5):1181–1187

Kirschmann DA, Seftor EA et al (2002) A molecular role for lysyl oxidase in breast cancer invasion. Cancer Res 62(15):4478–4483

Kitajima Y, Ide T et al (2008) Induction of hepatocyte growth factor activator gene expression under hypoxia activates the hepatocyte growth factor/c-Met system via hypoxia inducible factor-1 in pancreatic cancer. Cancer Sci 99(7):1341–1347

Korah R, Boots M et al (2004) Integrin alpha5beta1 promotes survival of growth-arrested breast cancer cells: an in vitro paradigm for breast cancer dormancy in bone marrow. Cancer Res 64(13):4514–4522

Koukourakis MI, Giatromanolaki A et al (2006) Comparison of metabolic pathways between cancer cells and stromal cells in colorectal carcinomas: a metabolic survival role for tumor-associated stroma. Cancer Res 66(2):632–637

Kroon ME, Koolwijk P et al (2000) Urokinase receptor expression on human microvascular endothelial cells is increased by hypoxia: implications for capillary-like tube formation in a fibrin matrix. Blood 96(8):2775–2783

Kuphal S, Bosserhoff AK (2006) Influence of the cytoplasmic domain of E-cadherin on endogenous N-cadherin expression in malignant melanoma. Oncogene 25(2):248–259

Lechner M, Lirk P et al (2005) Inducible nitric oxide synthase (iNOS) in tumor biology: the two sides of the same coin. Semin Cancer Biol 15(4):277–289

Lee WY, Chen HH et al (2005) Prognostic significance of co-expression of RON and MET receptors in node-negative breast cancer patients. Clin Cancer Res 11(6):2222–2228

Leek RD, Lewis CE et al (1996) Association of macrophage infiltration with angiogenesis and prognosis in invasive breast carcinoma. Cancer Res 56(20):4625–4629

Leong KG, Niessen K et al (2007) Jagged1-mediated notch activation induces epithelial-to-mesenchymal transition through Slug-induced repression of E-cadherin. J Exp Med 204(12):2935–2948

Lester RD, Jo M et al (2007) uPAR induces epithelial-mesenchymal transition in hypoxic breast cancer cells. J Cell Biol 178(3):425–436

Li G, Satyamoorthy K et al (2001) N-cadherin-mediated intercellular interactions promote survival and migration of melanoma cells. Cancer Res 61(9):3819–3825

Liao D, Corle C et al (2007) Hypoxia-inducible factor-1alpha is a key regulator of metastasis in a transgenic model of cancer initiation and progression. Cancer Res 67(2):563–572

Liao D, Luo Y et al (2009) Cancer associated fibroblasts promote tumor growth and metastasis by modulating the tumor immune microenvironment in a 4T1 murine breast cancer model. PLoS One 4(11):e7965

Lin EY, Nguyen AV et al (2001) Colony-stimulating factor 1 promotes progression of mammary tumors to malignancy. J Exp Med 193(6):727–740

Lin S, Sun L et al (2009) Chemokine C-X-C motif receptor 6 contributes to cell migration during hypoxia. Cancer Lett 279(1):108–117

Liu B, Fan Z et al (2009) Metformin induces unique biological and molecular responses in triple negative breast cancer cells. Cell Cycle 8(13):2031–2040

Mahon PC, Hirota K et al (2001) FIH-1: a novel protein that interacts with HIF-1alpha and VHL to mediate repression of HIF-1 transcriptional activity. Genes Dev 15(20):2675–2686

Maity A, Solomon D (2000) Both increased stability and transcription contribute to the induction of the urokinase plasminogen activator receptor (uPAR) message by hypoxia. Exp Cell Res 255(2):250–257

Martinez-Zaguilan R, Seftor EA et al (1996) Acidic pH enhances the invasive behavior of human melanoma cells. Clin Exp Metastasis 14(2):176–186

Maxwell PH, Wiesener MS et al (1999) The tumour suppressor protein VHL targets hypoxia-inducible factors for oxygen-dependent proteolysis. Nature 399(6733):271–275

Mazzone M, Dettori D et al (2009) Heterozygous deficiency of PHD2 restores tumor oxygenation and inhibits metastasis via endothelial normalization. Cell 136(5):839–851

Milosevic M, Fyles A et al (2001) Interstitial fluid pressure predicts survival in patients with cervix cancer independent of clinical prognostic factors and tumor oxygen measurements. Cancer Res 61(17):6400–6405

Mottet D, Castronovo V (2008) Histone deacetylases: target enzymes for cancer therapy. Clin Exp Metastasis 25(2):183–189

Muller A, Homey B et al (2001) Involvement of chemokine receptors in breast cancer metastasis. Nature 410(6824):50–56

Naidu R, Har YC et al (2009) Associations between hypoxia-inducible factor-1alpha (HIF-1alpha) gene polymorphisms and risk of developing breast cancer. Neoplasma 56(5):441–447

Niizeki H, Kobayashi M et al (2002) Hypoxia enhances the expression of autocrine motility factor and the motility of human pancreatic cancer cells. Br J Cancer 86(12):1914–1919

Oliveras-Ferraros C, Vazquez-Martin A et al (2009) Genome-wide inhibitory impact of the AMPK activator metformin on [kinesins, tubulins, histones, auroras and polo-like kinases] M-phase cell cycle genes in human breast cancer cells. Cell Cycle 8(10):1633–1636

Orimo A, Gupta PB et al (2005) Stromal fibroblasts present in invasive human breast carcinomas promote tumor growth and angiogenesis through elevated SDF-1/CXCL12 secretion. Cell 121(3):335–348

Pan Y, Mansfield KD et al (2007) Multiple factors affecting cellular redox status and energy metabolism modulate hypoxia-inducible factor prolyl hydroxylase activity in vivo and in vitro. Mol Cell Biol 27(3):912–925

Pennacchietti S, Michieli P et al (2003) Hypoxia promotes invasive growth by transcriptional activation of the met protooncogene. Cancer Cell 3(4):347–361

Petrella BL, Lohi J et al (2005) Identification of membrane type-1 matrix metalloproteinase as a target of hypoxia-inducible factor-2 alpha in von Hippel-Lindau renal cell carcinoma. Oncogene 24(6):1043–1052

Phoenix KN, Vumbaca F et al (2009) Therapeutic metformin/AMPK activation promotes the angiogenic phenotype in the ERalpha negative MDA-MB-435 breast cancer model. Breast Cancer Res Treat 113(1):101–111

Poellinger L, Johnson RS (2004) HIF-1 and hypoxic response: the plot thickens. Curr Opin Genet Dev 14(1):81–85

Putignani L, Raffa S et al (2008) Alteration of expression levels of the oxidative phosphorylation system (OXPHOS) in breast cancer cell mitochondria. Breast Cancer Res Treat 110(3):439–452

Qing G, Simon MC (2009) Hypoxia inducible factor-2alpha: a critical mediator of aggressive tumor phenotypes. Curr Opin Genet Dev 19(1):60–66

Rankin EB, Rha J et al (2008) Hypoxia-inducible factor-2 regulates vascular tumorigenesis in mice. Oncogene 27(40):5354–5358

Rauh MJ, Ho V et al (2005) SHIP represses the generation of alternatively activated macrophages. Immunity 23(4):361–374

Rius J, Guma M et al (2008) NF-kappaB links innate immunity to the hypoxic response through transcriptional regulation of HIF-1alpha. Nature 453(7196):807–811

Rizki A, Weaver VM et al (2008) A human breast cell model of preinvasive to invasive transition. Cancer Res 68(5):1378–1387

Rofstad EK, Tunheim SH et al (2002) Pulmonary and lymph node metastasis is associated with primary tumor interstitial fluid pressure in human melanoma xenografts. Cancer Res 62(3):661–664

Ruas JL, Poellinger L (2005) Hypoxia-dependent activation of HIF into a transcriptional regulator. Semin Cell Dev Biol 16(4–5):514–522

Sahlgren C, Gustafsson MV et al (2008) Notch signaling mediates hypoxia-induced tumor cell migration and invasion. Proc Natl Acad Sci USA 105(17):6392–6397

Scatena R, Bottoni P et al (2008) Glycolytic enzyme inhibitors in cancer treatment. Expert Opin Investig Drugs 17(10):1533–1545

Schietke RE, Warnecke C et al (2010) The lysyl oxidases LOX and LOXL2 are necessary and sufficient to repress E-cadherin in hypoxia – insights into cellular transformation processes mediated by HIF-1. J Biol Chem 26;285(9):6658–6669

Schindl M, Schoppmann SF et al (2002) Overexpression of hypoxia-inducible factor 1alpha is associated with an unfavorable prognosis in lymph node-positive breast cancer. Clin Cancer Res 8(6):1831–1837

Schioppa T, Uranchimeg B et al (2003) Regulation of the chemokine receptor CXCR4 by hypoxia. J Exp Med 198(9):1391–1402

Seagroves TN, Ryan HE et al (2001) Transcription factor HIF-1 is a necessary mediator of the pasteur effect in mammalian cells. Mol Cell Biol 21(10):3436–3444

Selak MA, Armour SM et al (2005) Succinate links TCA cycle dysfunction to oncogenesis by inhibiting HIF-alpha prolyl hydroxylase. Cancer Cell 7(1):77–85

Semenza GL (2003) Targeting HIF-1 for cancer therapy. Nat Rev Cancer 3(10):721–732

Serrati S, Margheri F et al (2008) Endothelial cells and normal breast epithelial cells enhance invasion of breast carcinoma cells by CXCR-4-dependent up-regulation of urokinase-type plasminogen activator receptor (uPAR, CD87) expression. J Pathol 214(5):545–554

Shackelford DB, Vasquez DS et al (2009) mTOR and HIF-1alpha-mediated tumor metabolism in an LKB1 mouse model of Peutz-Jeghers syndrome. Proc Natl Acad Sci USA 106(27): 11137–11142

Song G, Ouyang G et al (2009) Osteopontin promotes gastric cancer metastasis by augmenting cell survival and invasion through Akt-mediated HIF-1alpha up-regulation and MMP9 activation. J Cell Mol Med 13(8B):1706–1718

Spangenberg C, Lausch EU et al (2006) ERBB2-mediated transcriptional up-regulation of the alpha5beta1 integrin fibronectin receptor promotes tumor cell survival under adverse conditions. Cancer Res 66(7):3715–3725

Staller P, Sulitkova J et al (2003) Chemokine receptor CXCR4 downregulated by von Hippel-Lindau tumour suppressor pVHL. Nature 425(6955):307–311

Stockmann C, Doedens A et al (2008) Deletion of vascular endothelial growth factor in myeloid cells accelerates tumorigenesis. Nature 456(7223):814–818

Swinnen JV, Beckers A et al (2005) Mimicry of a cellular low energy status blocks tumor cell anabolism and suppresses the malignant phenotype. Cancer Res 65(6):2441–2448

Takeda N, O'Dea EL et al (2010) Differential activation and antagonistic function of HIF-{alpha} isoforms in macrophages are essential for NO homeostasis. Genes Dev 24(5):491–501

Tan EY, Campo L et al (2007) Cytoplasmic location of factor-inhibiting hypoxia-inducible factor is associated with an enhanced hypoxic response and a shorter survival in invasive breast cancer. Breast Cancer Res 9(6):R89

Tanimoto K, Makino Y et al (2000) Mechanism of regulation of the hypoxia-inducible factor-1 alpha by the von Hippel-Lindau tumor suppressor protein. EMBO J 19(16):4298–4309

Tarasenko N, Nudelman A et al (2008) Histone deacetylase inhibitors: the anticancer, antimetastatic and antiangiogenic activities of AN-7 are superior to those of the clinically tested AN-9 (Pivanex). Clin Exp Metastasis 25(7):703–716

Tatemichi M, Ogura T et al (2009) Impact of inducible nitric oxide synthase gene on tumor progression. Eur J Cancer Prev 18(1):1–8

Tatum JL, Kelloff GJ et al (2006) Hypoxia: importance in tumor biology, noninvasive measurement by imaging, and value of its measurement in the management of cancer therapy. Int J Radiat Biol 82(10):699–757

Tennant DA, Frezza C et al (2009) Reactivating HIF prolyl hydroxylases under hypoxia results in metabolic catastrophe and cell death. Oncogene 28(45):4009–4021

Thangasamy A, Rogge J et al (2009) Recepteur d'origine nantais tyrosine kinase is a direct target of hypoxia-inducible factor-1alpha-mediated invasion of breast carcinoma cells. J Biol Chem 284(21):14001–14010

Thiery JP (2002) Epithelial-mesenchymal transitions in tumour progression. Nat Rev Cancer 2(6):442–454

Treins C, Murdaca J et al (2006) AMPK activation inhibits the expression of HIF-1alpha induced by insulin and IGF-1. Biochem Biophys Res Commun 342(4):1197–1202

Tsutsui S, Yasuda K et al (2005) Macrophage infiltration and its prognostic implications in breast cancer: the relationship with VEGF expression and microvessel density. Oncol Rep 14(2): 425–431

Tsutsumi S, Yanagawa T et al (2004) Autocrine motility factor signaling enhances pancreatic cancer metastasis. Clin Cancer Res 10(22):7775–7784

Ullah MS, Davies AJ et al (2006) The plasma membrane lactate transporter MCT4, but not MCT1, is up-regulated by hypoxia through a HIF-1alpha-dependent mechanism. J Biol Chem 281(14):9030–9037

van der Groep P, Bouter A et al (2008) High frequency of HIF-1alpha overexpression in BRCA1 related breast cancer. Breast Cancer Res Treat 111(3):475–480

Vaupel P, Kallinowski F et al (1990) Blood flow, oxygen consumption and tissue oxygenation of human tumors. Adv Exp Med Biol 277:895–905

Vaupel P, Briest S et al (2002) Hypoxia in breast cancer: pathogenesis, characterization and biological/therapeutic implications. Wien Med Wochenschr 152(13–14):334–342

Vaupel P, Mayer A et al (2005) Hypoxia in breast cancer: role of blood flow, oxygen diffusion distances, and anemia in the development of oxygen depletion. Adv Exp Med Biol 566: 333–342

Vazquez-Martin A, Oliveras-Ferraros C et al (2009) The antidiabetic drug metformin suppresses HER2 (erbB-2) oncoprotein overexpression via inhibition of the mTOR effector p70S6K1 in human breast carcinoma cells. Cell Cycle 8(1):88–96

Wang GL, Jiang BH et al (1995) Hypoxia-inducible factor 1 is a basic-helix-loop-helix-PAS heterodimer regulated by cellular O2 tension. Proc Natl Acad Sci USA 92(12):5510–5514

Weigert A, Brune B (2008) Nitric oxide, apoptosis and macrophage polarization during tumor progression. Nitric Oxide 19(2):95–102

Xie K, Dong Z et al (1996) Activation of nitric oxide synthase gene for inhibition of cancer metastasis. J Leukoc Biol 59(6):797–803

Yan M, Rayoo M et al (2009) BRCA1 tumours correlate with a HIF-1alpha phenotype and have a poor prognosis through modulation of hydroxylase enzyme profile expression. Br J Cancer 101(7):1168–1174

Yang J, Mani SA et al (2004) Twist, a master regulator of morphogenesis, plays an essential role in tumor metastasis. Cell 117(7):927–939

Yang AD, Camp ER et al (2006) Vascular endothelial growth factor receptor-1 activation mediates epithelial to mesenchymal transition in human pancreatic carcinoma cells. Cancer Res 66 (1):46–51

Young SD, Hill RP (1990) Effects of reoxygenation on cells from hypoxic regions of solid tumors: anticancer drug sensitivity and metastatic potential. J Natl Cancer Inst 82(5):371–380

Young SD, Marshall RS et al (1988) Hypoxia induces DNA overreplication and enhances metastatic potential of murine tumor cells. Proc Natl Acad Sci USA 85(24):9533–9537

Yun H, Lee M et al (2005) Glucose deprivation increases mRNA stability of vascular endothelial growth factor through activation of AMP-activated protein kinase in DU145 prostate carci-noma. J Biol Chem 280(11):9963–9972

Zelzer E, Levy Y et al (1998) Insulin induces transcription of target genes through the hypoxia-inducible factor HIF-1alpha/ARNT. EMBO J 17(17):5085–5094

Zhong H, De Marzo AM et al (1999) Overexpression of hypoxia-inducible factor 1alpha in common human cancers and their metastases. Cancer Res 59(22):5830–5835

Zhuang Y, Miskimins WK (2008) Cell cycle arrest in Metformin treated breast cancer cells involves activation of AMPK, downregulation of cyclin D1, and requires p27Kip1 or p21Cip1. J Mol Signal 3:18

Zinser GM, Leonis MA et al (2006) Mammary-specific Ron receptor overexpression induces highly metastatic mammary tumors associated with beta-catenin activation. Cancer Res 66(24):11967–11974

Index